Managing Product Innovation

The views of five BNAC members:
Carrol Bolen, Ken Durham, Walter Light,
Chester Sadlow and Viscount Weir

BRITISH-NORTH AMERICAN COMMITTEE

Sponsored by
British-North American Research Association (U.K.)
National Planning Association (U.S.A.)
C. D. Howe Institute (Canada)

1030: La Teste-de-Buch (Gironde).
1032: Grande Lande.
1034: Dax (Landes).
1035: Aire-sur-l'Adour (Landes).
1036: Landes.
1044: Agen (Lot-et-Garonne).
1045: Lot-et-Garonne.
1048: Gers.
1057: Bagnères-de-Luchon (Haute-Garonne).
1065: Couserans (Ariège).
1068: Val d'Aran (Catalonia).
1078: Lavedan (Hautes-Pyrénées).
1080: Arréns (Hautes-Pyrénées).
1088: Lescar (Basses-Pyrénées).
1090, 1093, 1094: Béarn.
1098: Eaux-Bonnes (Basses-Pyrénées).
1101: Bayonne (Basses-Pyrénées).

98 WORDS AND DESCRIPTIVE TERMS FOR 'WOMAN' AND 'GIRL'

871: La Salle-Saint-Pierre (Gard).
873: Nîmes (Gard), Colognac (Gard).
877: Grau de Palavas (Hérault).
878: Aniane (Hérault).
880: Pézenas (Hérault).
881: Béziers (Hérault).
890: Carcassonne (Aude), Narbonne (Aude).
891, 891a: Lézignan (Aude).
892: Carcassonne (Aude).
895–897: Lauragues (Aude).
900: Pépieux (Aude).
901, 902: Region between Catalan frontier and Languedocian, in Pyrénées-Orientales and Aude.
906: Pyrénées françaises dialects.
915, 916: Toulousain speech of Haute-Garonne.
920: Tarn.
921: Castres (Tarn).
928: Quercy (Tarn-et-Garonne).
936: Vallée de la Bonnette (Lot).
939: Rouergue.
947: Rochessauve (Ardèche).
949: Gilhoc (Ardèche).
950: Vivarais (Ardèche).
956: Mons-la-Tour (Haute-Loire).
960: Carlat (Cantal).
962: Aurillac (Cantal).
967: Ytrac (Cantal)
968: Laroquebrou (Cantal).
969: Saint Simon (Cantal).
970: Mauriac (Cantal).
972: Patois used from Murat as far as Molompise (Cantal).
974: Puy-de-Dôme.
975: Basse-Auvergne.
977: Auvergne.
978: Vinzelles (Puy-de-Dôme).
980: Ambert (Puy-de-Dôme).
982: Limagne patois, esp. of Riom (Puy-de-Dôme).
985: Creuse.
987: Chavanat (Creuse).
988: Aubusson (Creuse).
989: Benevent (Creuse).
990: Southern part of Guéret (Creuse).
991: Tulle (Corrèze).
997: Limousin.
1002: Puybarraud (Charente).
1004, 1008, 1009: Périgord.
1015: Saint-Pierre-de-Chignac (Dordogne).
1018, 1019: Villeneuve-sur-Lot (Lot-et-Garonne).
1020, 1022, 1023: Gascogne.
1028: Pyrénées françaises.

729: Vaux-en-Bugey (Ain).
731: Bresse.
733, 734, 743, 748: Lyon (Rhône).
742: Lyon (Rhône), Forez (Loire), Beaujolais (Rhône).
744, 755: Saint-Genis-les-Ollières (Rhône).
745: Grand'Côte.
749: Couzon (Rhône).
750, 751: Villefranche-sur-Saône (Rhône).
752: Létra (Rhône).
753, 754: Grézieu (Rhône).
756, 759: Saint-Etienne (Loire).
758: Forez (Loire).
760: Saint-Haon-le-Châtel (Loire).
763, 767: Dauphiné.
764: Isère.
771: Jons (Isère).
772: Crémieu (Isère).
773: Terres-Froides (Isère).
775: Grenoble (Isère).
777: Voiron (Isère).
778: Beaurepaire (Isère).
780: Saint-Maurice-de-l'Exil (Isère).
783: Gascogne.
786: Nice to Bayonne, from Pyrénées to central France.
788: Languedoc, Provence, Gascogne, Béarn, Quercy, Rouergue, Limousin, Bas-Limousin, Dauphiné.
789: Provençal.
795: Midi de la France.
798: Mens (Isère).
800: Hautes-Alpes, Basses-Alpes.
803: Alpes du Dauphiné.
805: Briançonnais and vallées Vaudoises of Alpes Cottiennes, esp. Queyras (Hautes-Alpes).
806: Hautes-Alpes, esp. Gap and Champsaur.
807: Champsaur (Hautes-Alpes).
808: Bruis (Hautes-Alpes) and Vallée de l'Oule (Hautes-Alpes).
809: Gap (Hautes-Alpes).
810: Lallé (Hautes-Alpes).
811: Drôme.
813: Die (Drôme).
814: Charpey (Drôme).
815: Vaudois (Piémont).
816: Pragelato (Piémont).
825: Roaschia (Piémont).
828, 837: Provençal.
848: Aix (Bouches-du-Rhône).
849, 851: Marseille.
858: Vallée de Barcelonnette (Basses-Alpes).
862: Nice (Alpes-Maritimes).
865, 866, 868, 870: Languedoc.

617: Crémine (Berne).
620: Montbéliard (Doubs).
622: Bournois (Doubs).
623: Baume-les-Dames (Doubs).
624: Franche-Montagne patois, esp. Damprichard (Doubs).
625: La Grand'combe (Doubs).
627: Fourgs (Doubs).
628: Sancey (Doubs), Mesnay (Jura), Vitteaux (Côte-d'Or).
629: Besançon (Doubs), Vanclans (Doubs).
632: Jura.
634: Dôle (Jura).
637: Salins (Jura).
638: Jura.
640: Crans (Jura).
643: Provençal and Franco-Provençal dialects.
647: Suisse Romande dialects, esp. Alpes Vaudoises and Val d'Illiez.
651: French Swiss dialects.
654: Suisse Romande.
660, 661: Suisse romane dialects.
667: Neuchâtel (Switz.).
668: Fribourg (Switz.).
669: Dompierre (Fribourg).
670: Guyère (Fribourg).
672: Vaud (Switz.).
673: Vallée de Joux (Vaud), Mouthe (Doubs), Bois d'Amont (Jura), Chénit (Vaud).
674: Jorat (Vaud).
675: Blonay (Vaud).
676: Alpes Vaudoises.
677: Vionnaz (Bas-Valais).
678: Val d'Illiez (Valais).
679: Val de Bagnes (Valais).
682–684: Savièse (Valais).
685: Grimisaut (Valais).
686: Montana (Valais).
687: Italian and South Swiss Linguistic Atlas.
698: Val Soana (Piémont).
701: Usseglio (Piémont).
702: Faeto (Italy, Foggia), Celle (Italy, Foggia).
703: Savoie region.
709: Certoux (Genève).
711: Chapelle d'Abondance (Haute-Savoie).
712: Magland (Haute-Savoie).
713: Albertville (Savoie).
715: Tarentaise (Savoie).
716: Bonneval (Savoie).
720: Saint-Lupicin (Jura).
721: Poisoux (Jura).
723: Bresse and Bugey regions, esp. Valromey (Ain).
725: Jujurieux (Ain).
726: Cerdon (Ain).

481: Messon (Aube).
486: Ramerupt (Aube).
486*bis*: Méry-sur-Seine (Aube).
493: Esternay (Marne).
495: Essarts-lez-Sézanne (Marne).
500: Vavray (Marne).
501: Possesse (Marne).
502, 503: Courtisols (Marne).
508: Château-Thierry (Aisne).
512: Ardennes.
518: Buzancy (Ardennes).
522: Mouzonnais, esp. Bulson (Ardennes).
532: Lorrain patois.
540: Brillon (Meuse), Chattancourt (Meuse), Peuvillers (Meuse), Vignot (Meuse).
541: Dombras (Meuse).
542: Mangiennes (Meuse).
543: Seuzey (Meuse).
546: Tannois (Meuse).
548: Maxey-sur-Vaise (Meuse).
552, 553: Gaumet in Wallon region.
555: Prouvy (Nord), Jamogne (Luxembourg).
558: Lorrain patois of Metz and Belfort along German linguistic frontier.
560: Moselle.
570: Rémilly (Moselle).
572: Fauquemont (Moselle).
574: Hattigny (Moselle), Ommeray (Moselle).
575, 578, 580: Vosges.
581: Fiménil (Vosges).
579: Uriménil (Vosges).
583: Ban-de-la-Roche (Bas-Rhin), Lunéville (Meurthe-et-Moselle).
584: La Baroche (Haut-Rhin), Belmont (Bas-Rhin).
586: La Baroche (Haut-Rhin).
588: Southwestern Vosges dépt.
589: Remiremont (Vosges).
590–592: Southern Vosges region.
594: Aillevillers (Haute-Saône), Val d'Ajol (Vosges), Plombières (Vosges), Les Granges (Vosges), Fougerolles (Haute-Saône), Saint-Bresson (Haute-Saône), Cleurie (Vosges).
597: Dommartin (Vosges).
598: Saint-Amé (Vosges).
600: Franche-Comté, Lorraine, Alsace.
601: Franche-Comté.
604: Pierrecourt (Haute-Saône).
607: Broye-lez-Pesmes (Haute-Saône).
608: Luxeuil (Haute-Saône).
609: Plancher-les-Mines (Haute-Saône).
611: Belfort territory.
612: Vagney (Vosges).
613: Trouée de Belfort.
614: Chatenois (Belfort) and other localities in Belfort and Jura Bernois territory.
616: Sornetan (Berne).

305: Pléchâtel (Ille-et-Vilaine).
307: Pipriac (Ille-et-Vilaine).
309: Basse-Bretagne.
310, 311: Quimper (Finistère).
319: Maine.
320: Bas-Maine.
330: Anjou and esp. Segré (Maine-et-Loire).
334, 342: Poitou.
349: Vendée.
353: Ile d'Elle (Vendée).
356: Bas-Gâtinais (Deux-Sèvres).
357: Lezay (Deux-Sèvres).
359: Niort (Deux-Sèvres).
360: Chef-Boutonne (Deux-Sèvres).
364: Saintonge.
370: Charente-Inférieure.
373: Seudre valley (Charente-Inférieure), Seugne valley (Charente-Inférieure).
375: Saint-Amant (Charente).
383: Loches (Indre-et-Loire).
384: Touraine.
388: Blois (Loir-et-Cher).
394: Gâtinais (Loiret).
395: Bonneval (Eure-et-Loir).
397: Berry.
398, 399, 401: Centre de la France region including Cher, Indre, Indre-et-Loire, and Nièvre.
411: Nohant (Indre), Saint-Chartier (Indre), La Châtre (Indre).
414: Charost (Cher).
419: Bourbonnais.
423: Montluçon (Allier).
425: Ferrières (Allier).
426: Bourgogne.
433: Morvan.
434, 435: Chaulgnes (Nièvre).
443: Clessé (Saône-et-Loire), Mâcon (Saône-et-Loire).
445: Bresse-Louhannaise (Saône-et-Loire).
446: Montret (Saône-et-Loire).
447: Bresse-Châlonnaise (Saône-et-Loire), Saint-Germain-du-Bois (Saône-et-Loire).
448: Verduno-chalonnais in Saône-et-Loire and northwest Ain region.
450: Côte-d'Or.
452: Dijon (Côte-d'Or).
453: Minot (Côte-d'Or).
454: Sainte-Sabine (Côte-d'Or).
459: Bourberain (Côte d'Or).
467: Yonne.
468: Thory (Yonne), Avallon (Yonne).
469: Châtel-Censoir (Yonne).
474: Vermenton (Yonne).
477: Ligny (Yonne), Seignelay (Yonne).
478: Champagne.
480: Bercenay-en-Othe (Aube).

BIBLIOGRAPHY

146, 147, 150, 151: Namur.
154: Boninne-lez-Namur (Namur).
158: Chimay (Hainaut), Andenne (Namur).
160: Perwez (Brabant).
164: Mons (Hainaut).
169: Bray (Hainaut), Papignies (Hainaut).
171: Marches-lez-Ecaussines (Hainaut).
175: Flandre.
176: Lille (Nord).
182: Gondecourt (Nord).
183: Douai (Nord).
186: Rouchi (Nord).
189: Nord.
198: Boulogne-sur-Mer (Pas-de-Calais).
202: Saint-Omer (Pas-de-Calais).
203: Créquy (Pas-de-Calais), Fressin (Pas-de-Calais), Sains (Pas-de-Calais), Torcy (Pas-de-Calais).
204: Saint-Pol (Pas-de-Calais).
208: Vraignes (Somme).
209: Démuin (Somme).
211: Picardie.
212: Friedrichsdorf (am Taunus) (Hesse-Nassau).
214, 217, 222: Normandy.
219: Lisieux (Calvados).
226: Seine-Inférieure.
233: Le Havre (Seine-Inférieure).
235: Pont-Audemer (Eure).
239: Ezy-sur-Eure (Eure).
241: Thaon (Calvados).
244: Bayeux (Calvados).
252: Bocage normand (Calvados).
254: Bessin (Calvados).
263: Orne.
264, 265: Alençon (Orne).
269: Orne and bordering regions of Mayenne and La Sarthe.
275: Cherbourg (Manche), Saint-Lô (Manche), Valognes (Manche).
276: Manche.
279: Cotentin (Manche), Port-Bail (Manche), Saint-Sauveur-le-Vicomte (Manche).
282: Gréville (Manche), La Hague (Manche).
284: Guernesey.
285: Jersey.
286: Aurigny (Iles Normandes).
287: Bretagne.
288: Ille-et-Vilaine.
290: Ille-et-Vilaine and Manche.
294: Dol (Ille-et-Vilaine).
296: Cogles (Ille-et-Vilaine).
299, 300: Rennes (Ille-et-Vilaine).
302: Ercé-près-Liffré (Ille-et-Vilaine).
303: Gennes-sur-Seiche (Ille-et-Vilaine).
304: Gosné (Ille-et-Vilaine), Saint-Aubin (Ille-et-Vilaine).

Littré, Maximilien Paul Emile: Dictionnaire de la langue française, 4 vols.; Paris, 1873.
Métivier, Georges: Dictionnaire franco-normand, vi + 499 pp.; London and Edinburgh, 1870.
Meyer-Lübke, Wilhelm: Romanisches etymologisches Wörterbuch[3], xxiii + 1204 pp.; Heidelberg, 1935.
Moll, Francisco de B., transl.: C. H. Grandgent, Introducción al Latín vulgar, 384 pp.; Madrid, 1928.
Murray, Sir James A. H., with others: A New English Dictionary on Historical Principles, 10 vols. and supplement; Oxford, 1888–1933.
Passy, Paul, and George Hempl: International French-English and English-French Dictionary, 637 pp.; New York, 1904.
Pauli, Ivan, 'Enfant', 'Garçon', 'Fille', dans les langues romanes; Lund, 1919.
Sainéan, L.: Les sources indigènes de l'étymologie française, 3 vols.; Paris, 1930.
Wartburg, Walther von: Bibliographie des dictionnaires patois, 146 pp.; Paris, 1934.

The numbers below refer to Wartburg's bibliographical items, and the names to the respective localities explored:

2: Bouillon, Carcassonne, L'Armagnac, Mont-de-Marsan, Perigourdin, Saint-Omer.
3, 4, 5: Atlas Linguistique de la France.
6: Französisches Etymologisches Wörterbuch.
21: French dialects in general.
25, 26, 29: Paris.
38: Verviers (Liège).
39: Namur.
40, 41, 44, 55, 59, 63, 64, 66–77, 116, 118–121, 141: Wallon.
57: Malmédy (Liège), Spa (Liège), Verviers (Liège), Liège.
58: Ardenne region, Condroz (Dinant), Hesbaye (Liège), Herve (Liège).
81: Malmédy, Francorchamps (Liège).
82: Malmédy.
83: Faymonville (Malmédy).
84: Gueuzaine (Malmédy), Waimes (Malmédy).
88–90: Verviers (Liège).
91: Stavelot (Liège).
92: Coo (Belgium, Stavelot).
93, 96 Spa (Liège).
104, 105, 111, 117: Liège.
123, 124: Jupille (Liège).
126: Ans (Belgium, Liège).
128: Fléron (Liège), Romsée (Liège).
127: Vottem (Liège).
131: Seraing (Liège).
133, 137: Luxembourg.
134: Vielsalm (Luxembourg).
136: Saint-Hubert (Luxembourg).
138: Cherain (Luxembourg).
139: Nassogne (Luxembourg).
142: Ardenne region.
143: Bouillon (Luxembourg).
145: Givet (Ardennes).

BIBLIOGRAPHY

Adams, Edward Larrabee: Word Formation in Provençal xvii + 607 pp.; New York, 1913.
Adams, George C. S., and Clement M. Woodard: A Census of French and Provençal Dialect Dictionaries in American Libraries, 17 pp.; Linguistic Society of America, Lancaster, Pa., 1937.
Augé, Claude, ed.: Nouveau Petit Larousse Illustré, 1760 pp.; Paris, 1934.
Baker, Ernest A., and Frank H. Vizetelly, ed.: Cassell's New French and English Dictionary, 1020 pp.; New York, 1930.
Bloch, Oscar: Dictionnaire étymologique de la langue française (avec la collaboration de W. von Wartburg, 2 vols.; Paris, 1932.
Brütting, J., Das Bauern-Französisch in Dancourts Lustspielen; Erlangen, 1911.
Cotgrave, R.: A Dictionarie of the French and English Tongues; London, 1611.
Dauzat, Albert: Dictionnaire étymologique de la langue française, xxxvii + 762 pp.; Paris, 1928.
D'Hauterive, R. Grandsaignes: Dictionnaire d'Ancien Français, xi + 592 pp.; Paris, 1947.
Diez, Friedrich Christian: Etymologisches Wörterbuch der romanischen Sprachen[5], xxvi + 866 pp.; Bonn, 1887.
Diez, Friedrich Christian: Grammatik der romanischen Sprachen[5], xix + 1134 pp.; Bonn, 1882.
Dinneen, P. S.: An Irish-English Dictionary, xxx + 1340 pp.; Dublin, 1927.
Ernout, Alfred, and Antoine Meillet, Dictionnaire étymologique de la langue latine; Paris, 1932.
Fuller, Thomas: History of Cambridge and Waltham Abbey, xxiv + 688 pp.; London, 1840.
Gamillscheg, Ernst: Etymologisches Wörterbuch der französischen Sprache, xxvi + 1136 pp.; Heidelberg, 1928.
Gilliéron, Jules Louis: Généalogie des mots qui designent l'abeille, 360 pp.; Paris, 1918.
Godefroy, Frédéric: Dictionnaire de l'ancienne langue française et de tous ses dialectes du IXe au XVe siècle, 10 vols.; Paris, 1881–1903.
Godefroy, Frédéric: Lexique de l'ancien français, 544 pp.; Paris, 1901.
Grandgent, Charles Hall: An Introduction to Vulgar Latin, xvii + 219 pp.; Boston, 1907.
Grandgent, Charles Hall: An Outline of the Phonology and Morphology of Old Provençal, xi + 159 pp.; Boston, 1905.
Hatzfeld, Adolphe, Arsène Darmesteter and Antoine Thomas, Dictionnaire général de la langue française du commencement du XVIIe siècle à nos jours[9], Paris, 1927.
Heath's New French and English Dictionary, xxiv + 582 pp.; New York, 1932.
Holmes, Urban Tigner, and Alexander H. Schutz: A History of the French Language, vii + 184 pp.; New York, 1938.
Jarník, Johann Urban: Index zu Diez' etymologischen Wörterbuch der romanischen Sprachen, x + 378 pp.; Heilbronn, 1889.
Körting, Gustav: Lateinisch-romanisches Wörterbuch[3], vi + 787 pp.; Paderborn, 1907.

CHAPTER XXVIII

CONCLUSION

The distribution for *domina* and *femina* ia as follows: *dam, damne, dem, fom, fan, fam* (N. E. France and German border); *dame, fam, femme* (N. France and Belgian border); *dam, danme, dem, damma, dâma, fena, fana* (E. Central France and Swiss border); *donne, dana, dama, damo, femo, fenne, femno, fremo* (S. E. France and Italian border); *damo, dono, done, douno, fenno, henno, femne* (S. W. France and Spanish border); *dame, fam, femme* (Central France); *dame, femme* (W. Central France); *dame, fam, femme* (N. W. France).

A large majority of the descriptive terms considered are of Latin origin. Many words lost their original meanings and came to take on pejorative meanings. However, in at least half the cases considered, the original and assumed meanings were both kept (depending largely on the regions where used). The descriptive terms fall into a number of groups. In the greatest number of instances unseemly behavior is designated. Next in importance are terms denoting unattractive physical appearance. The two types of meaning are often found in combination. More than half of the terms derive their significance from the extension of meaning of terms applied to animals (horses, dogs, etc.), unmentionable parts of the body, and inanimate objects of various sorts. There are also a fair number of terms used primarily as injurious epithets.

That nearly all of the descriptive terms are of a derogatory nature is due to the fact that there is a general tendency, when speaking ill of a person, to sum him up in a word; whereas when one speaks well of a person, he usually makes use of locutions rather than single words.

vessiere 'insult applied to market women and which is explained by the word which follows' (226; Seine-Inférieure).
vielho 'old woman' (1008; Périgord).
virâgot 'virago, mannish woman' (759; Loire).
vivre 'resolute young girl' (450; Côte-d'Or).
vŏ 'woman who is easily led' (204; Saint-Pol).
voraina 'depraved, rascally libertine' (647; Suisse Romande).
voreinta 'depraved, rascally libertine' (647; Suisse Romande).
vouipa 'wasp; shrew' (647; Suisse Romande).
vouivre 'shrew' (632; Jura).
vuippa 'wasp; evil-tongued woman' (647; Suisse Romande).
wande 'lazy, dirty woman' (2, 143; Bouillon).
wape 'wasp; shrew' (552; Lorraine).
wench 'prostitute' (217; Anglo-Norman).
wèrai 'insulting name given to a woman' (560; Moselle).
whore 'prostitute' (217; Anglo-Norman).
yimpe 'abandoned woman' (373; Saintonge).
zabai 'woman of bad repute' (81; Malmédy, Wallon).
zaguel'resse 'streetwalker' (83; Faymonville).
zaguete 'woman of bad repute' (81; Malmédy, Wallon).
zaubiate, -ote 'Elizabeth; foolish, silly woman' (560; Moselle).
zèzèle 'stupid, idiotic woman' (560; Moselle).
zobette 'young girl who has a sweetheart' (622; Franche-Comté).

tranleto 'girl of bad life' (828; Provence).
tranliasso 'badly groomed streetwalker' (828; Provence).
tranliasso 'woman of bad life' (789; Provençal).
traquoire 'scatterbrained girl' (745; Rhône).
trawire 'negligent, lazy woman' (532; Allain).
trekayê 'woman dressed with affectation' (675; Vaud).
trênêa 'woman without care' (353; Vendée, Ile-d'Elle).
trenodo 'dishonest woman' (1002; Limousin, Périgord).
tribatt 'unnatural woman, pervert' (105; Liège).
trine 'girl, little girl; young prostitute' (81; Malmédy, Wallon).
trinnèle 'negligent, lazy woman' (560; Moselle).
trivaine 'streetwalker' (609; Franche-Comté).
troïe 'sow; public woman' (38; Wallon).
trolote 'prattler' (560; Moselle).
tronche 'fat woman or girl' (667; Suisse Romande).
tronchenette, trontchenette 'fat little woman' (667; Suisse Romande).
trôpe 'dirty servant' (632; Jura).
trosse-pète 'term of scorn for a little girl' (560; Moselle).
trouille 'dirty woman, slut' (601; Franche-Comté).
troūleūse 'woman who neglects her house in order to loaf' (560; Moselle).
trouille 'very fat, untidy woman' (595; Eure-et-Loir).
trouillon 'fat, dirty charwoman' (777; Dauphiné).
troute 'shameless, licentious woman' (105; Liège).
troue, truy 'slattern, fat slovenly woman; slut' (667; Suisse Romande).
troyo 'sow; prostitute' (980; Puy-de-Dôme).
truche 'annoying woman' (1093; Béarn).
truegeasso 'very dirty woman' (828; Provence).
truegeo 'dirty woman' (828; Provence).
trufle 'giddy, hairbrained woman' (234; Guernesey).
trūye 'sow; untidy glutton' (560; Moselle).
tschouma 'old she-ass' (647; Suisse Romande).
tsikkala 'little girl' (647; Suisse Romande).
tšofa 'overweight woman who eats all the time' (675; Vaud).
tsotta 'awkward, dirty, foolish girl' (647; Suisse Romande).
tšufa 'badly dressed, ruffled-haired woman' (675; Vaud).
tšurla 'little girl who weeps while yelling' (675; Vaud).
tuluoar 'neglected woman' (182; Gondecourt).
turchon 'duster; dirty woman' (622; Franche-Comté).
turlucane 'very innocent, unattractive old maid' (330; Anjou).
tutute 'prostitute' (105; Liège).
tutuuyl 'woman without order' (182; Gondecourt).
tyura 'woman who has her head upset' (675; Vaud).
vache 'cow; lazy woman' (759; Loire).
valariasso 'good girl who loves to laugh' (789; Provençal).
varloupa 'prostitute' 632; Jura).
vastibousière 'dirty woman, servant of low estate' (214; Normandie).
vedèle 'woman who plays the fool, one who is finical' (353; Vendée, Ile-d'Elle).
veille 'old woman' (356; Deux-Sèvres).
vergasse 'abashed woman' (419; Maine).
versa 'girl with a high bust escaping from her bodice' (647; Suisse Romande).

tamponne 'dirty woman' (632; Jura).
tantêlă 'very shallow woman' (703; Savoie).
tappa 'garrulous woman' (647; Suisse Romande).
taque 'prattler' (667; Suisse Romande).
tarale 'light, giddy woman' (214; Normandie).
tardarasso 'ragged woman of the scum of the people' (828; Provence).
tarote 'prattler' (522; Champagne, Ardennes).
tarouge 'endless talker' (143; Bouillon).
tarougi 'endless talker' (143; Bouillon).
tatille 'prattler' (486; Champagne, Aube).
tatillon 'prattler' (486; Champagne, Aube).
tatouille 'untidy, babbling woman' (399; Berry).
taudion 'sloven' (579; Vosges).
taudion 'dirty, disheveled woman' (560; Moselle).
taunique 'insipid woman' (214; Normandie).
tatouye 'careless housekeeper, prattler' (560; Moselle).
taudion 'dowdy woman' (532; Mailly).
tavasse 'unbearable prattler' (353; Vendée, Ile-d'Elle).
teupa 'heavy woman' (577; Suisse Romande).
tevène 'insulting term for woman' (373; Saintonge).
timbourletto 'plump little woman' (789; Provençal).
tirassado 'whore, streetwalker' (828; Provence).
titè-poupée 'courtezan' (851; Marseille).
titi 'woman to be scorned' (667; Suisse Romande).
tocson 'coarse woman' (214; Normandie).
toka 'heavy, stupid woman' (677; Suisse Romande).
tolion 'dirty woman, slut' (632; Jura).
tomale 'nonchalant woman' (81; Malmédy, Wallon).
topa 'naive, weak woman' (686; Suisse Romande).
tosa 'young girl' (866; Languedoc).
tougne 'indolent woman' (443; Clessé-en-Mâconnais).
touillon 'disordered, dirty woman' (522; Champagne, Ardennes).
touina 'soft, fat woman; sort of jacket' (810; Hautes-Alpes).
toûlass 'fat woman, glutton' (105; Liège).
toupî 'slut, streetwalker' (38; Wallon).
toupīe 'wicked, shrewish woman' (560; Moselle).
toupie 'light woman' (330; Anjou).
toupies 'woman of bad life' (226; Seine-Inférieure).
tourniresse 'woman without deportment or conduct, who, instead of working, turns from one thing to another' (214; Normandie).
toutelle 'woman without order' (176; Lille).
toutouye 'squalid little girl' (171; Marche-lez-Ecaussines).
toutusso 'silly woman' (789; Provençal).
trach 'disorderly, slovenly, licentious woman' (105; Liège).
traignière 'public girl' (384; Touraine).
trainahia 'prostitute' (647; Suisse Romande).
trainat 'woman of bad life' (330; Anjou).
traineau 'dirty woman' (356; Deux-Sèvres).
trakotyauza 'knitter' (675; Vaud).
tralé 'disorderly, lazy woman' (675; Vaud).
traméné 'woman of bad conduct' (675; Vaud).

WORDS AND DESCRIPTIVE TERMS FOR 'WOMAN' AND 'GIRL'

sagane 'term of scorn when applied to a woman' (1100; Gascogne).
sakatrap 'old witch' (105; Liège).
salo-toupi 'dolt' (865; Provençal).
sampanna 'careless woman' (373; Saintonge).
sampano 'untidy woman' (983; Puy-de-Dôme).
sampion 'woman of bad life' (982; Puy-de-Dôme).
sampo 'untidy woman' (982; Puy-de-Dôme).
sansouille 'disordered, dirty woman' (442; Bourgogne).
sardrouille 'dirty woman' (632; Jura).
sargail 'fickle woman' (356; Deux-Sèvres).
sargalle 'young girl careless about her person' (395; Eure-et-Loir).
sargounello 'untidy woman' (980; Puy-de-Dôme).
sarpiassi 'restless woman' (373; Saintonge).
sarrazino 'Saracen; woman having immodest allures' (805; Dauphiné).
satrelle 'thin-legged woman who jumps when she walks' (622; Franche-Comté).
sautegouillot 'untidy woman' (419; Maine).
sauterelle 'thin-legged woman who jumps when she walks' (622; Franche-Comté).
savetaillon 'woman dressed without taste' (330; Anjou).
sawoureûss 'saucy, impertinent woman; précieuse' (105; Liège).
schranké 'bad woman' (758; Suisse Romande).
séleri 'celery; insult addressed to an old woman' (622; Franche-Comté).
sempiternelle 'very old woman' (828; Provence).
sergale 'giddy woman' (214; Normandie).
siguenne 'cigogne, insulting name for a woman' (622; Franche-Comté).
sihla 'angry little girl who squeals like a little pig' (675; Vaud).
simplasso 'good girl' (789; Provençal).
sirpo 'wicked woman' (789; Provençal).
sitreûte 'ridiculously affected woman' (105; Liège).
skïriou 'girl of light morals' (703; Savoio).
sogrëne 'woman lacking in care' (353; Vendée, Ile-d'Elle).
soka 'very stupid woman' (686; Suisse Romande).
soudadiero 'streetwalker of low degree, prostitute of the barracks' (786; Nice, Bayonne).
soudarerasse, -osse 'bold, often shameless woman' (560; Moselle).
souirasso 'unclean woman' (789; Provençal).
souiro 'woman of bad life' (789; Provençal).
souldadas 'tall, shameless woman' (921; Tarn).
soulyèn 'untidy woman' (353; Vendée, Ile-d'Elle).
sourrastro 'bad sister' (789; Provençal).
sousternery 'wet nurse' (643; Franco-Provençal, Provençal).
strumpet 'prostitute' (217; Anglo-Norman).
sucotte 'ugly, talkative old woman' (622; Franche-Comté).
sucrado 'sweet; woman who affects to be modest, innocent scrupulous' (828; Provence).
suegro 'old woman who spies on other women' (828; Provence).
sumio 'monkey; affected woman' (805; Dauphiné).
surlurette 'decided woman' (269; Normandie, Mancelle).
tabure 'constant talker' (143; Bouillon).
tabuse 'loquacious woman' (675; Vaud).
tagas 'dirty, careless woman' (865; Dauphiné).
tagas 'dirty, careless woman' (789; Provençal).

pretantaine 'coquette, light woman' (226; Seine-Inférieure).
profemena 'prude, honest woman' (837; Vaucluse).
prosērpine 'clever, restless woman; termagant; she-devil' (560; Moselle).
prouf 'premature young girl, grotesque woman' (105; Liège).
puggy 'little girl, term of tenderness' (217; Anglo-Norman).
pupta 'woman of bad life' (686; Suisse Romande).
puselière 'woman who has fleas' (745; Rhône).
pusla 'whore' (217; Anglo-Norman).
puss 'whore' (217; Anglo-Norman).
puzzle 'whore' (217; Anglo-Norman).
pyorna 'woman constantly complaining' (675; Vaud).
quean 'bad woman' (217; Anglo-Norman).
quegne 'old doll, woman badly rigged who lives a disorderly life' (667; Suisse Romande).
queûrîe 'corpse, woman of bad ways' (552; Lorraine).
quincorne 'slow, irresolute old woman' (777; Dauphiné).
quinque 'sickly, complaining woman' (667; Suisse Romande).
racasse 'prattler' (384; Touraine).
radoueiri 'woman of evil ways' (373; Saintonge).
ragota 'little, dumpy girl' (712; Savoie).
raguette 'bold, quarrelsome woman' (143, 2; Bouillon).
rakrocheûss 'thieving streetwalker' (105; Liège).
rapporteuse 'woman who relates what one does and says' (38; Wallon).
rattes-penades 'women who wore rats in their wigs' (330; Anjou).
ravenelle 'lively, petulant woman' (164; Wallon).
rebalado 'strumpet' (865; Provençal).
rebaldo 'woman of bad life' (786; Nice, Bayonne).
rébecca 'chatterer, one who retorts freely' (522; Champagne).
recauqueto 'woman lacking warmth or animation' (789; Provençal).
redàfia 'woman of easy morals' (758; Haute-Savoie).
redassa 'thrush; thin woman' (647; Suisse Romande).
relicite, relite 'widow' (667; Suisse Romande).
ribandelle 'debauchee, lecherous woman' (226; Seine-Inférieure).
rig 'tomboyish and lively girl' (217; Anglo-Norman).
rioche 'woman who laughs without reason' (353; Vendée, Ile-d'Elle).
ripace 'pretentious woman' (353; Vendée, Ile-d'Elle).
ripo 'worthless man or woman' (814; Drôme).
ritchardo 'rich woman' (1002; Limousin, Périgord).
rodrigue 'evil-minded old woman' (419; Maine).
rondiche 'plump girl' (398; Centre).
ronion 'fat, massive woman' (217; Anglo-Norman).
ross 'shameless slattern' (105; Liège).
roubiaco 'melancholy old woman' (789; Provençal, Limousin).
roucouye 'woman of loose morals' (560; Moselle).
rouwaye 'debauchee' (560; Moselle).
rôzir 'virtuous young girl, a girl who had obtained the price of goodness in a town' (38; Wallon).
rutasse 'woman of bad life' (299; Rennes).
rutasse 'woman of bad life' (632; Jura).
ržola 'woman who is always laughing' (712; Savoie).
sado 'old and bad woman' (214; Normandie).

84 WORDS AND DESCRIPTIVE TERMS FOR 'WOMAN' AND 'GIRL'

piée 'woman of bad life' (330; Anjou).
pignasse 'scolding woman' (299; Rennes).
pignasse 'habitual scolder' (632; Jura).
pigno-bregâoudo 'wasp; shrewish, unbearable woman' (987; Marche).
pignorchi 'delicate woman, one of difficult taste' (373; Saintonge).
pîncêle 'lady-bird' (560; Moselle).
pinèguète 'restless, delicate, puny little girl' (560; Moselle).
pingelle 'gay, attractive little girl' (622; Franche-Comté).
piorna 'annoying, scolding complainer' (647; Suisse Romande).
pioule 'lazy, nonchalant, complaining woman' (703; Savoie).
pissouse 'woman, little girl' (373; Saintonge).
pitoueto 'young girl, plump and jolly girl' (786; Nice, Bayonne).
plâco 'talker' (987; Marche).
pleuche 'good-for-nothing, nonchalant, headstrong girl' (399; Berry).
pllanta-lezi 'idle news-monger' (647; Suisse Romande).
pocasse 'little girl' (393; Loiret).
poertse 'dirty woman' (677; Suisse Romande).
poew 'skin; woman of bad life' (1002; Limousin, Périgord).
pofo 'plump, fresh-looking girl' (991; Limousin, Périgord).
poiëtt 'damsel' (105; Liège).
poison 'woman of bad life' (330; Anjou).
popioûl 'affected woman, prude' (105; Liège).
porchi 'arrogant, pretentious, disdainful woman' (759; Loire).
pouaina 'woman with a malignant, pointed face' (647; Suisse Romande).
pouairtza 'slut, dirty woman, shameless girl' (647; Suisse Romande).
pouelle 'young girl' (393; Loiret).
poueso 'prostitute' (789; Provençal).
pouésous 'abandoned woman' (375; Charente).
pouffiasse 'fat woman; shameless woman' (419; Maine).
pouffiasse 'fat, dirty woman, and even worse' (748; Lyon).
pouffiasse 'fat, flabby woman' (745; Rhône).
pouffiasse 'light woman' (356; Deux-Sèvres, Bas-Gâtinais).
poufiasse 'woman of bad life' (1094; Béarn).
poufiasse 'prostitute' (151; Wallon-Français).
pouhouye 'dirty woman' (138; Cherain).
pouina 'woman with a malignant, pointed face' (647; Suisse Romande).
poulotte 'young girl' (447; Saône-et-Loire).
poume-pouère 'abandoned girl who is neither maiden, wife, nor widow (398; Centre).
poundeyro 'hen that gives many eggs; prolific woman (921; Tarn).
poupardiere 'woman with large breasts' (1094; Béarn).
poupardiero 'woman with a large bosom' (789; Provençal).
poupardo 'fleshy woman' (789; Provençal).
poupāye 'doll; young girl' (560; Moselle).
poupude 'woman with large breasts' (1094; Béarn).
poūse 'wife' (560; Moselle).
pôvress 'poor woman, beggar' (38; Wallon).
poyssardo 'vulgar woman, woman of the scum of the people, of the markets' (921; Tarn).
predere 'loose woman, woman of bad life' (1094; Béarn).
prenhe 'pregnant woman; animal with young' (1093; Béarn).
prèsèrpine 'restless woman' (560; Moselle).

nouna 'foolish girl' (715; Savoie).
nounǎ 'slow, stupid woman' (703; Savoie).
novi 'bride, fiancée' (789; Provençal).
nula 'meddler' (703; Savoie).
ofrega 'woman dressed badly and with no taste' (712; Savoie).
ortchie 'lively tireless old woman' (622; Franche-Comté).
ouipa 'wasp; shrew' (647; Suisse Romande).
ouivette 'dizzy young girl' (214; Normandie).
ourjas 'woman without order' (789; Provençal).
ouyate 'little goose; foolish woman' (560; Moselle).
pachaou 'indolent female' (851; Marseille).
paioulado 'woman in child-bed' (789; Provençal).
palache 'fat woman with large hands; tall, dried-up woman' (560; Moselle).
palate 'tall, dried-up woman' (560; Moselle).
palhourado 'woman in child-bed' (789; Provençal).
pallongue 'dirty, reviled, poor old woman' (777; Dauphiné).
pano 'untidy woman' (982; Puy-de-Dôme).
panouche 'rags; woman in tatters; woman without morals' (789; Provençal).
pantragno 'dirty woman' (789; Provençal).
panturla 'dirty woman; tall, thin person' (858; Basses-Alpes).
panturlo 'woman of bad life' (838; Vaucluse).
paourouno 'young beggar' (828; Provence).
parraule 'fat woman' (1094; Béarn).
particulière 'woman of ill repute' (29; Paris).
passeuse 'In Lyon, women who pass people on the river Saône' (643; Franco-Provençal, Provençal).
patace 'woman slow in her work' (703; Savoie).
pathrolie 'dirty woman' (715; Savoie).
patholla 'garrulous woman' (647; Suisse Romande).
pebrino 'wicked shrew' (1023; Gascogne).
pedas 'swaddling clothes; debauchee' (789; Provençal).
pekina 'foolish, disagreeable blockhead' (647; Suisse Romande).
pekka 'foolish, disagreeable blockhead' (647; Suisse Romande).
pelhas 'rags; woman in tatters; woman without morals' (789; Provençal).
pella 'sow; bad woman' (759; Loire).
penauza 'hairdresser, coiffeuse' (675; Vaud).
pèque 'bad woman of biting tongue, rather stupid' (522; Champagne, Ardennes).
percho 'pole; tall, thin woman' (828; Provence).
péroine 'talker' (300; Rennes).
pesque 'term of insult; proud, harsh woman' (226; Seine-Inférieure).
pétasse 'woman or girl; prostitute' (667; Suisse Romande).
pétasse 'narrow-minded woman' (353; Vendée, Ile-d'Elle).
petasse 'woman attached to nonsense' (300; Rennes).
petasse 'babbler' (632; Jura).
petite 'little girl' (29; Paris).
peto-bas 'term of scorn applied to females' (789; Provençal).
petonton 'woman attached to nonsense' (300; Rennes).
petoufias 'plump or fat woman' (789; Provençal).
piccolina 'young girl' (758; Jura).
picêle 'young girl who is going to sing trimazaus' (560; Moselle).
pièce 'cracked, infatuated woman' (399; Berry).

miglelenne 'one who squints' (622; Franche-Comté).
mighia 'married woman' (758; Jura).
mignate 'young girl' (560; Moselle).
mignouna 'pet, favorite' (779; Dauphiné).
mihou 'woman of bad conduct' (560; Moselle).
mille 'women' (758; Jura).
mimuetche 'ridiculous, affected woman' (622; Franche-Comté).
mineresse 'minor' (540; Meuse).
minot 'girl' (589; Vosges).
miorle 'clumsy woman' (1094; Béarn).
mioura 'indolent, lazy, slovenly woman' (777; Dauphiné).
mique 'silly, awkward, tall, thin girl without charm' (748; Lyon).
miss 'miss' (217; Anglo-Norman).
monéri 'young lady of honor' (712; Savoie).
molessa 'woman, female' (445; Saône-et-Loire).
molle 'cock; nonchalant woman' (662; Franche-Comté).
mone 'sullen woman' (560; Moselle).
monîn 'fat, ugly woman' (560; Moselle).
monine 'bad woman' (745; Rhône).
monnon 'foolish, sullen, disagreeable girl' (647; Suisse Romande).
morio 'little girl' (813; Drôme).
morlate 'streetwalker, young girl who frequents boys' (560; Moselle).
morveluse 'snooty, glandered woman' (760; Loire).
morveuse 'gamin, urchin' (226; Seine-Inférieure).
moteintze 'accursed woman, witch' (647; Suisse Romande).
motré 'dirty, ruffle-haired woman' (560; Moselle).
mougnardo 'scowling woman' (789; Provençal).
moujasse 'very young girl' (356; Deux-Sèvres).
mouna 'ugly woman, old cow' (373; Saintonge).
moune 'ugly woman of bad life; monkey' (1094; Béarn).
moungeto 'little nun' (789; Provençal).
mounino 'ugly woman; monkey' (786; Nice, Bayonne).
mousette 'frolicsome, roguish little girl' (233; le Havre).
moyus 'coquettish young girl' (298; Bretagne).
moza 'young girl; heifer' (647; Suisse Romande).
mugnate 'marriageable young girl' (614; Franche-Comté).
mugnate 'young girl of marriageable age' (613; Alsace).
mulasso 'sterile woman, woman of bad life; old mule' (1023; Gascogne).
myauna 'boring, complaining woman' (675; Vaud).
naneta 'limping woman' (675; Vaud).
nanon 'simple woman' (675; Vaud).
naulière 'news-monger' (275; Normandie, Manche).
niaugne 'narrow-minded, shallow woman' (674; Puy-de-Dôme).
nigōn 'woman with little intelligence' (182; Gondecourt).
nino 'little girl' (789; Provençal).
niqudolïe 'badly dressed, dirty woman' (703; Savoie).
noce 'shallow, awkward woman' (703; Savoie).
noka 'slow, dull woman' (675; Vaud).
nono 'nun; grandmother; little girl' (789; Provençal).
noouvio 'bride' (813; Drôme).
nouna 'silly, awkward girl' (647; Suisse Romande).

mare 'woman easily led' (204; Saint-Pol).
mare 'unworthy mother; mother of an animal' (675; Vaud).
mareȳn 'godmother, woman' (307; Ille-et-Vilaine, Pipriac).
margalo 'streetwalker' (814; Drôme).
margaude 'a woman who is nothing' (748; Lyon).
margaude 'woman of bad life' (745; Rhône).
margelle 'giddy girl' (601; Franche-Comté).
margot 'magpie; woman drunkard' (745; Rhône).
margot 'daisy; woman who is worthless' (789; Provençal).
margot 'woman of bad life' (980; Puy-de-Dôme).
margot 'daisy; woman of bad conduct' (233; le Havre).
margoton 'woman of bad life' (745; Rhône).
margouton 'magpie; talkative woman; woman of equivocal morals' (759; Loire).
marie-bon gas 'girl of bad life' (384; Touraine).
marie-brasoc 'woman who never leaves the corner of the hearth' (1094; Béarn).
marigraillon 'dirty, ill-dressed girl; slut' (647; Suisse Romande).
marioar 'woman who seeks marriage' (182; Gondecourt).
markange 'girl of medium virtue' (647; Suisse Romande).
marrerasse 'mayor's wife, wealthy woman' (560; Moselle).
masco 'mask, false face; old, ugly woman' (921; Tarn).
massipo 'young girl who lives in a *mas*' (786; Nice, Bayonne).
mastroulhe 'fat, ugly woman' (1094; Béarn).
matokka 'silly, clumsy, graceless girl' (647; Suisse Romande).
matres 'mistress' (594; Vosges).
matrônn 'matron, grave, elderly woman' (105; Liège).
matroquette 'sharp, crafty, sly young woman' (164; Wallon).
matta 'little girl, simple little girl, doll' (647; Suisse Romande).
maynade 'child, young girl' (1093; Béarn).
māzète 'giddy little girl; one who does not grow; cow' (560; Moselle).
mèdjalate 'young girl' (560; Moselle).
meg 'tall woman; pole' (217; Anglo-Norman).
megie 'abbreviation for Marguerite; stupid young girl' (622; Franche-Comté).
mègneye 'young girl of the house; mistress; sweetheart, fiancée; servant' (560; Moselle).
megnotta 'little girl' (647; Suisse Romande).
meigco-fremo 'woman lacking intelligence' (828; Provence).
meinadello 'pretty woman' (789; Provençal).
meirino 'oldest woman in the household' (789; Provençal).
menaijer 'woman or girl' (296; Bregagne, Coglais-Français).
mendigo 'young girl' (789; Provençal).
mendio 'girl; young girl seeking a *mari mendiga*' (805; Dauphiné).
menin 'careful woman' (789; Provençal).
menesse 'woman' (29; Paris).
meniurro 'liar; charlatan' (1023; Gascogne).
menoille 'woman disorderly in appearance' (300; Rennes).
méraude 'woman of bad life, who has children' (353; Vendée, Ile-d'Elle).
merdgelle 'giddy girl' (601; Franche-Comté).
merluche 'tall, dried-up woman' (745; Rhône).
meune 'decrepit old woman' (83; Faymonville).
mèyāye 'wife' (560; Moselle).
mèyon 'Mary; woman of bad life' (560; Moselle).

leuva 'wanton, debauchee' (677; Suisse Romande).
liērpi, niērpi 'bunch of grapes from which the grapes are gone; shrewish, wicked, loud woman' (991; Limousin, Périgord).
limpa 'light woman' (758; Savoie).
logne 'nonchalant, indolent, heedless woman' (164; Wallon).
lone 'prattler, driveler' (622; Franche-Comté).
lostière 'dupe, libertine' (164; Wallon).
loueiri 'woman of equivocal ways' (373; Saintonge).
loueiri 'debauched woman' (756; Loire).
louiêri 'prostitute, woman for hire' (759; Loire).
louiriano 'woman lacking in sense and deportment' (789; Provençal).
lou-tro 'completely abandoned woman of the worst kind' (991; Limousin, Périgord).
louiro 'prostitute' (789; Provençal).
luronna 'strong woman, virago' (647; Suisse Romande).
luskette 'woman who squints' (2, 143; Bouillon).
macaque 'ugly woman, old ugly woman; woman of bad life' (1094; Béarn).
macipe 'girl, servant' (1093; Béarn).
maclas 'an abandoned woman' (764; Dauphiné, Isère).
maclas, maclassi 'libertine who imitates actions of boys and seeks their company' (763; Dauphiné).
macorre 'woman of bad life' (1094; Béarn).
macrale 'witch; hypocrite' (151; Wallon-Français).
madouneto 'woman of low condition' (789; Provençal, Toulouse).
magaw 'toothless old woman' (105; Liège).
magnin 'badly dressed, ill-famed woman' (522; Champagne).
magrite 'shrew' (151; Wallon-Français).
maguilone 'little girl, brat, girl of ill repute' (81; Malmédy, Wallon).
mahule 'badly dressed, ill-formed woman; wicked woman' (560; Moselle).
mainado 'young girl' (786; Nice, Bayonne).
makako 'monkey; strumpet' (1023; Gascogne).
makapoîe 'slut, lazy woman' (105; Liège).
makareou 'prostitute' (1023; Gascogne).
makrott 'streetwalker, procuress' (105; Liège).
malaisée 'woman's sobriquet applied as a joke' (399; Berry).
mâlaud 'virago, vixen, man-like woman' (398; Centre).
malla 'young girl' (686; Suisse Romande).
mâl-magritt 'shrew' (105; Liège).
mâlot 'unreserved, masculine girl' (411; Berry, Nohant).
mamaie 'prostitute' (105; Liège).
māmiche 'grandmother; old woman' (560; Moselle).
mancipo 'young girl' (789; Provençal).
mandourre 'woman who has a dull mind' (1094; Béarn).
mandro 'beggar, procuress' (786; Nice, Bayonne).
mandro 'fox; sly woman' (881; Hérault).
mandrouno 'procuress' (838; Vaucluse).
maniago 'affected, spoiled young woman' (786; Nice, Bayonne).
manion 'debauchee' (212; Saint Omer).
mano 'old ewe; sterile woman' (789; Provençal, Gascogne).
maon 'woman' (81; Malmédy, Wallon).
maprovaie 'evil-minded, shameless, unruly woman' (105; Liège).
maraoudo 'slut; scoundrel' (1023; Gascogne).

jhacasse 'busybody' (373; Saintonge).
jhan-fenno 'dolt' (865; Provençal).
jhavéră 'prattler' (703; Savoie).
jigandeina 'woman who has bad carriage' (858; Basses-Alpes).
jorso 'streetwalker' (789; Provençal).
le joujour, mon jour 'a very ugly woman' (411; Berry, Nohant).
joventas 'young girls' (928; Quercy).
kanl 'prattler' (136; Saint-Hubert).
kânôie 'lazy, sleepy-headed woman' (38; Wallon).
kapetsé 'woman with a difficult walk' (675; Vaud).
karkevala 'girl who runs awkwardly' (675; Vaud).
karkevallha 'talkative woman, gossip' (647; Suisse Romande).
katola 'sickly woman' (647; Suisse Romande).
katoza 'squalid woman' (677; Suisse Romande).
kayena 'very dirty woman' (675; Vaud).
kègne 'slut, bitch; girl of bad life; lazy person' (560; Moselle).
kêmarê 'parasite' (677; Suisse Romande).
kètîn 'slut, sweetheart, lover' (560; Moselle).
kicksey-wicksey 'silly weak woman' (217; Anglo-Norman).
kigeâzeuse 'disparager, slanderer' (38; Wallon).
kôka 'wicked old woman' (647; Suisse Romande).
kontkubina 'concubine' (675; Vaud).
kopineûss 'talker, prattler' (105; Liège).
kopinress 'talker, prattler' (105; Liège).
koreûss 'streetwalker' (105; Liège).
koûrress 'streetwalker' (105; Liège).
kovase 'woman who uses the *covet*' (675; Vaud).
krapota 'young girl fifteen to sixteen years old' (712; Savoie).
krapôtt 'girl, little girl, sweetheart' (38; Wallon).
krokanna 'wicked woman' (647; Suisse Romande).
krokia 'wicked old woman' (647; Suisse Romande).
kunantine 'woman or animal constantly running' (1002; Limousin, Périgord).
kura 'silly, credulous young girl' (647; Suisse Romande).
kwila 'bawling woman' (675; Vaud).
lacai 'offspring; affected woman, précieuse' (789; Provençal).
laideronne 'ugly creature' (667; Suisse Romande).
laidouno 'ugly though not unattractive woman' (828; Provence).
laidron 'ugly woman' (735; Lyon).
laisse 'ugly woman' (330; Anjou).
lanbruchë 'tall, shallow, languid woman' (703; Savoie).
landra 'dissipated girls' (764; Dauphiné).
landre 'woman' (758; Lorraine).
landroie 'slovenly girl or woman' (532; Le Tholy).
landroye 'negligent, lazy woman' (560; Moselle).
landrôye 'careless, non-alert woman' (579; Vosges).
landrusa 'woman' (758; Usseglio).
lânnress 'thief, cheat' (105; Liège).
lanterno 'lantern'; *vielho lanterno* 'hideous old woman' (789; Provençal).
larica 'great talker' (763; Dauphiné).
laye 'woman who neglects her person and her household' (560; Moselle).
lebriero 'greyhound; streetwalker' (789; Provençal).

havasse 'dissolute woman, sullen woman' (57; Wallon).
havet 'dirty woman' (214; Normandie).
hawi 'idiot woman, innocent woman' (57; Wallon).
helegaud 'tall, ill-formed woman' (57; Wallon).
heniho 'half-crazy woman who walks the streets' (57; Wallon).
héridelle 'young girl' (284; Guernesey).
hérote 'gawky woman' (560; Moselle).
hèrpeute 'worn-out instrument; soft, languid person; giddy, light girl; sort of oath' (560; Moselle).
hervai 'potsherd; old woman no longer good for anything' (57; Wallon).
hervete 'talkative woman, hussy' (57; Wallon).
hignoteuss 'titterer, sneerer' (57; Wallon).
hinque 'woman with disproportionate and thin figure' (57; Wallon).
hirlaha 'woman who chatters and is indiscreet' (57; Wallon).
hitate 'badly reared little girl' (57; Wallon).
hoirneuse 'shifty, evading woman' (57; Wallon).
holeuse 'woman who brews mischief, is irresolute and hesitant' (57; Wallon).
homias 'women' (1020; Gascogne).
hore 'prostitute' (217; Anglo-Norman).
hosse-cowe 'hoche-queue; woman who makes difficulties' (57; Wallon).
hotte 'old woman' (594; Vosges).
houhou 'owl; dirty woman' (560; Moselle).
houlette 'prostitute' 284; Guernesey).
houleuse 'one who weeps, groans aloud' (57; Wallon).
houlotte 'serious owl; young girl, last-born girl of family' (57; Wallon).
houlpineresse 'nonchalant, idle woman' (57; Wallon).
houpetata 'heedless young girl' (57; Wallon).
houprale 'screech owl; slut, disorderly woman' (57; Wallon).
houstus 'distracted woman' (214; Normandie).
houvion 'map; dirty woman' (57; Wallon).
houzar 'soldier; virago, femme gardie' (57; Wallon).
huppe 'woman careless of her person' (300; Bretagne, Rennes).
hurluberlu 'brusk and brutal woman' (57; Wallon).
imbrouniasso 'habitual drunkard' (991; Limousin, Périgord).
inglitin 'red herring; thin, emaciated woman' (57; Wallon).
intremetteuse 'procuress, proxeneta' (57; Wallon).
ispusa 'spouse' (701; Piemont).
jabiâssi 'thrush; chattering woman' (759; Loire).
jabra 'ungainly woman' (398; Centre).
jabra 'ungainly woman' (397; Berry).
jagouasse 'lightning celandine; prattler' (398; Centre).
jaguilene 'silly, imbecile woman' (57; Wallon).
jaillo 'blond woman' (805; Dauphiné).
jairieuse 'envious one who covets all she sees' (57; Wallon).
jappa 'scolding talebearer' (647; Suisse Romande).
jappette 'cur; woman who talks much but is not wicked' (233; le Havre).
jasasse 'gossip' (214; Normandie).
jatéro 'young girl who runs after boys' (1023; Gascogne).
javasse 'busybody' (356; Deux-Sèvres).
javatte 'name of scorn given to a woman' (477; Côte-d'Or).
jeunesse 'young girl or young man' (397; Berry).

MISCELLANY 77

guessa 'woman' (758; Jura).
guet(r)o 'gaiter; woman of bad life' (789; Provençal).
gueuie 'mouth; bad and slandering woman' (57; Wallon).
gueusa 'pig iron; bad woman' (759; Loire).
gueuymate 'libertine' (560; Moselle).
guezette 'dizzy, insolent girl' (214; Normandie).
guiaulette 'woman who weeps often' (233; le Havre).
guibre 'nag; wicked woman' (233; le Havre).
guiene 'bad housekeeper' (57; Wallon).
guigne 'woman of bad and dissolute ways' (57; Wallon).
guingete 'little girl who plays young lady' (81; Malmédy, Wallon).
guina 'ragged woman' (715; Savoie).
guinche 'badly dressed, ill-famed woman' (522; Champagne).
guinche 'woman to be scorned' (373; Saintonge).
guindeino 'badly dressed woman' (789; Provençal).
guinguete 'little girl who plays the young lady' (57; Wallon).
guiniôche 'badly dressed, dirty woman; woman of bad life' (703; Savoie).
guirouleja 'young girl who has the mannerisms of a boy' (1094; Béarn).
guiroulhe 'young girl who has the mannerisms of a boy' (1094; Béarn).
guso 'kept woman; sow' (1023; Gascogne).
habaja 'woman who speaks at random' (57; Wallon).
hacha 'gossip, noisy woman' (57; Wallon).
hacha 'chatterer, fussy woman' (38; Wallon).
hag 'old witch' (217; Anglo-Norman).
hagaie 'ungainly woman' (57; Wallon).
hagnante 'woman who retorts spitefully' (57; Wallon).
haguette 'gossip; foolish, dizzy woman' (57; Wallon).
haie 'young wench, person one can not get rid of' (57; Wallon).
halbouieuse 'woman who haggles, wavers' (57; Wallon).
halebaie 'dowdy, awkward girl' (57; Wallon).
halkineu 'one who wavers, evader, loiterer' (38; Wallon).
halkineuse 'hesitant woman' (57; Wallon).
haloche 'woman who carries herself poorly' (143, 2; Bouillon).
haloppe 'badly dressed woman' (2; Bouillon).
hamande 'constant complainer' (560; Moselle).
hanne 'old woman' (264, 214; Normandie, Orne, Alençon).
haquenayo 'ungainly, awkward woman' (828; Provence).
harboie 'squalid woman' (57; Wallon).
hardeie 'toothless old lady' (57; Wallon).
hardelle 'ugly, very thin woman' (26; Paris).
hardelle 'complaisant young girl' (214; Normandie).
harèjress 'fish-woman; coarse, indolent, quarrelsome woman' (105; Liège).
harengire 'bad woman' (143; Bouillon).
haridan 'bad horse; debauched woman' (217; Anglo-Norman).
haridelle 'frivolous person' (57; Wallon).
harikrute 'little shell; unaccomplished woman' (57; Wallon).
harlot 'prostitute' (217; Anglo-Norman).
harotte 'bad horse; woman of no merit' (57; Wallon).
harpie 'harpy; noisy and wicked woman' (828; Provence).
harrias 'difficult woman' (330; Anjou).
hase, vieille hase 'old rabbit; old woman' (233; le Havre).

76 WORDS AND DESCRIPTIVE TERMS FOR 'WOMAN' AND 'GIRL'

gougouie 'woman who loves a good time' (57; Wallon).
goulimando 'public girl' (786; Nice, Bayonne).
goulimando 'public woman' (921; Tarn).
goullamas 'dirty, idle woman' (881; Hérault).
goumai 'coarse woman, little fat woman' (105; Liège).
goumai 'fat and dumpy woman' (57; Wallon).
goumaie 'overcooked brick; indolent and thick-set woman' (57; Wallon).
goumaïe 'dumpy and indolent woman' (105; Liège).
goumha 'disgusting, dirty, ugly woman' (647; Suisse Romande).
goundino 'hussy; streetwalker' (921; Tarn).
gourdonine 'streetwalker' (353; Vendée, Ile-d'Elle).
goureusse 'woman who cheats in market or game' (57; Wallon).
gourgandine 'prostitute' (609; Franche-Comté).
gourgandine 'streetwalker in a bad section' (395; Eure-et-Loir, Bonneval).
gourgâoudo 'girl of bad life' (987; Marche).
gourrine 'woman of bad life' (745; Rhône).
goye 'girl of bad life' (560; Moselle).
goyote 'badly dressed woman' (453; Côte d'Or).
goyro 'buzzard, kite; woman whose hair is ruffled' (921; Tarn).
grabeusse 'crayfish; old woman' (579; Vosges).
graillon 'dirty woman who smells bad' (748; Lyon).
grandiveuse 'vain woman who affects to despise her neighbor' (57; Wallon).
grand'mée 'grandmother; old woman' (164; Wallon).
graniu 'fertile land; prolific woman' (982; Puy-de-Dôme).
grelotte 'public girl' (647; Suisse Romande).
grenadier 'tall, strong, tough-looking woman' (57; Wallon).
gribiche 'wicked old woman with whom to scare children' (214; Normandie).
griffarda 'lady' (758; Savoie).
grigwése 'shrewd, sharp, bold woman' (560; Moselle).
grigneusse 'grumbler' (57; Wallon).
grigoisse 'sly woman, canteen woman of a free and bold humor' (57; Wallon).
grindo 'streetwalker' (789; Provençal).
gringo 'girl of bad life' (789; Provençal).
gripe 'brusk, petulant, quick-tempered girl' (426; Bourgogne).
gripette 'wicked woman who may use her claws' (57; Wallon).
grippette 'cross, naughty girl' (176; Lille).
griwése 'bold, shrewd, sharp woman' (560; Moselle).
grognasse 'woman of ill repute' (29; Paris).
grolate 'grumbler' (579; Vosges).
grolâte 'scolding woman' (532; Rehaupal).
groumieuse 'toothless old dame who chews her gums' (57; Wallon).
groumiott 'decrepit, toothless old woman' (105; Liège).
groumott 'decrepit, toothless old woman' (105; Liège).
groumotte 'toothless and decrepit old woman' (57; Wallon).
growe 'woman of loose morals' (57; Wallon).
guega 'jade; foolish, difficult woman' (858; Basses-Alpes).
guèhhenīre 'girl who runs after boys' (560; Moselle).
guenon 'she-monkey; dirty woman' (38; Wallon).
guenoun 'monkey; very ugly woman of bad life' (828; Provence).
guenuche 'disgusting woman who attires herself with baubles, streetwalker' (38; Wallon).

MISCELLANY

gerlo 'jar; said of a girl who has lost her virginity' (789; Provençal).
germotte 'year-old sheep; puny little girl' (57; Wallon).
gialotte 'little devil' (622; Franche-Comté).
gigă 'woman of huge stature' (703; Savoie).
gigogne 'a woman having many children' (330; Anjou).
gigue 'organ of Barbary; giddy girl' (622; Franche-Comté).
giletta 'little girl' (807; Champsaur).
gipoutre 'awkward, mannish girl' (284; Guernesey).
girande 'woman in child-bed' (398; Centre).
girl 'girl' (217; Anglo-Norman).
girouio 'wild carrot; woman lacking self-assurance' (789; Provençal).
glawenne 'gossip, slanderer' (57; Wallon).
glen 'chicken; negligent, woman without energy' (204; Saint-Pol).
gliâoudo 'stupid, foolish woman' (987; Marche).
glotte 'glutton' (57; Wallon).
gniniolle 'dull, soft, careless woman' (622; Franche-Comté).
gnioche 'silly, foolish woman' (745; Rhône).
gnogne 'booby; a little fool' (789; Provençal).
gnognotte 'one who devotes much time to little things' (419; Maine).
gnola 'courtezan' (758; Alpes Piémontaises).
gnorle 'clumsy woman' (1094; Béarn).
gnougne 'stupid woman who wants always to be right' (748; Lyon).
gnoune 'silly, foolish woman' (745; Rhône).
gobeie 'malicious, bad woman' (81; Malmédy, Wallon).
gobeie 'little pig; wicked and malicious woman' (57; Wallon).
godalle 'woman vender of remedies to produce abortions' (57; Wallon).
gode 'dirty, ugly woman' (143, 2; Bouillon).
godelette 'woman who guzzles' (164; Wallon).
godineta 'public girl' (764; Dauphiné).
godo 'bad ewe; bad woman' (881; Hérault).
godo 'old ewe; lazy woman, woman of bad life' (789; Provençal).
gogne 'pad or bag which holds petticoats; untidy woman' (398; Centre).
golippe 'dirty woman of equivocal reputation' (143; Bouillon).
gondèn 'carefree woman' (353; Vendée, Ile-d'Elle).
gone 'badly dressed, scorned woman' (632; Jura).
gongon 'habitual scolder' (632; Jura).
gonnella 'girl who looks and acts as if she were foolish' (763; Dauphiné).
gonnelle 'dress; foolish-looking girl of light morals' (803; Dauphiné).
gonzo 'prostitute' (786; Nice, Bayonne).
gonzo 'streetwalker' (789; Provençal).
goodgings 'sockets of the helm; little females' (217; Anglo-Norman).
gopa 'fat woman' (675; Vaud).
gopa 'fat and robust woman, girl of medium virtue, slut' (647; Suisse Romande).
gôpe 'girl of bad conduct' (632; Jura).
gorgando 'streetwalker, woman liking a good time' (987; Marche).
gorja 'woman' (758; Alpes Piémontaises).
gossip 'godmother; gabbling woman, shrewd woman' (217; Anglo-Norman).
gouâlioun 'dirty girl' (987; Marche).
goudineto 'gay working-girl, coquette, little bourgeoise of bad morals' (786; Nice, Bayonne).
gougënye 'untidy woman' (353; Vendée, Ile-d'Elle).

ganeou 'robust, indefatigable woman' (858; Basses-Alpes).
ganipa 'woman without bearing of any sort' (806; Hautes-Alpes).
ganipo 'woman of bad life' (980; Puy-de-Dôme).
ganippo 'woman of bad life' (982; Puy-de-Dôme).
ganto 'wild goose; woman of bad life' (789; Provençal).
gaoupas 'slut, bad girl; shameless girl' (921; Tarn).
gâpeille 'spendthrift' (356; Deux-Sèvres).
garapă 'mannerless woman' (703; Savoie).
garauda 'shameless woman, prostitute' (647; Suisse Romande).
garaude 'woman of bad life' (745; Rhône).
gareio 'giddy woman; light woman' (789; Provençal).
garella 'sharp girl' (759; Loire).
gargousse 'prostitute' (57; Wallon).
gariâne 'girl of bad life' (419; Maine).
garille 'sow; woman of little value' (419; Maine).
garlandresse 'wasteful woman' (57; Wallon).
garnul 'woman lacking energy and character' (204; Saint-Pol).
garodo 'ragged, shameless woman' (982; Puy-de-Dôme).
garoulo 'woman lacking order' (789; Provençal).
garrouio 'young wood; dirty woman' (789; Provençal).
garzeliano 'girl of bad deportment' (789; Provençal).
gasawet 'little girl' (459; Côte-d'Or).
gaselle 'sow' (398; Centre).
gasteuse 'glutton' (57; Wallon).
gate 'girl' (419; Maine).
gateirou 'woman of bad life' (862; Alpes-Maritimes).
gatte 'goat; careless, dirty woman' (57; Wallon).
gaudineto 'woman of medium virtue' (789; Provençal).
gaudrille 'prostitute' (426; Bourgogne).
gaupe 'untidy woman' (397; Berry).
gaupe 'bad woman and worse' (748; Lyon).
gaupe 'slovenly, disagreeable woman' (991; Limousin, Périgord).
gaupe 'negligent woman' (426; Bourgogne).
gaurauda 'streetwalker' (373; Saintonge).
gaure 'carefree, fat woman' (214; Normandie).
gausserando 'debauchee' (789; Provençal).
gavaude 'careless woman' (419; Maine).
gawedieuse 'cautious, crafty woman' (57; Wallon).
gaye 'fat, jovial woman, shameless girl' (560; Moselle).
gaynole 'term of scorn applied to women' (1094; Béarn).
gazette 'gazette; news-monger, talkative woman' (622; Franche-Comté).
gazille 'girl; girl of medium virtue; girl seven or eight years old' (419; Maine).
geamouna 'badly dressed woman or girl' (858; Basses-Alpes).
géane 'giantess' (735; Lyon).
gemihate 'groaning woman' (57; Wallon).
genemetsé 'curious, indiscreet woman' (675; Vaud).
genil 'rags; dirty woman' (204; Saint-Pol).
gens 'strumpet; simple woman' (675; Vaud).
gens 'women' (142; Ardennes).
gentifemo 'noble woman' (789; Provençal).
geonnfeie 'unmarried girl' (38; Wallon).

MISCELLANY

frisket 'wide-awake young girl' (204; Saint-Pol).
fumreie 'married woman, widow, unmarried woman' (105; Liège).
furbec 'great talker' (789; Provençal).
furlangueûss 'lavish dissipater' (105; Liège).
gabion 'woman of difficult character and doubtful morals who keeps disagreeable company' (980; Puy-de-Dôme).
gabras 'powerful girl of free allures' (810; Hautes-Alpes).
gabre 'gander; shameless, impudent girl' (789; Provençal).
gabrello 'basket; woman who does not know how to dress' (789; Provençal).
gaburre 'great talker' (789; Provençal).
gadaoula 'woman of bad life' (858; Basses-Alpes).
gadin 'good thing; old woman' (57; Wallon).
gadrô 'woman without care' (353; Vendée, Ile-d'Elle).
gadroie 'soft meat; wicked old woman' (57; Wallon).
gadroue 'careless, dirty woman' (522; Champagne, Ardennes).
gadrouye 'woman without order' (171; Marche-les-Ecaussines).
gadyi 'foolish, untidy woman' (675; Vaud).
gafroūse 'woman badly dressed, with hair disheveled' (560; Moselle).
gãgâna 'noon-day bell; slow woman who is always late' (675; Vaud).
gahete 'woman who conceives, becomes pregnant quickly' (1094; Béarn).
gailoche 'woman who is nothing' (453; Côte d'Or).
gaioufas 'dirty woman' (789; Provençal).
gaisso 'bitch; dirty or debauched woman' (789; Provençal).
gaîte 'gay woman' (419; Maine).
galabre 'gander; shameless, impudent girl' (789; Provençal).
galanda 'frisky, alluring, pleasure-seeking girl' (647; Suisse Romande).
galehouse 'streetwalker' (81; Malmédy, Wallon).
galehouse 'public woman' (57; Wallon).
galipoto 'fantastic being with which one scares children; masked one who sows terror; woman of sordid appearance' (980; Puy-de-Dôme).
gallaise 'courtezan' (226; Seine-Inférieure).
galle 'bad and quarrelsome woman' (57; Wallon).
galoeŵ,- oez 'spiteful or wicked individual' (204; Saint-Pol).
galosé 'woman who washes often' (675; Vaud).
galosse 'slut, disorderly woman' (57; Wallon).
galzebre 'great talker' (789; Provençal).
gambie 'base, vile woman' (622; Franche-Comté).
gamelle 'sow; fat woman' (398; Centre).
gamiche 'girl of bad life' (632; Jura).
gampo, gaupo 'slut' (789; Provençal).
ganache 'blockhead' (198; Boulonnais).
ganbasse 'libertine' (703; Savoie).
gandalia 'streetwalker' (959; Auvergne).
gandarme 'manlike woman' (57; Wallon).
gandeina 'streetwalker' (858; Basses-Alpes).
gandeutha 'debauchee, wayward woman' (678; Suisse Romande).
gandeutha 'sinful woman, debauchee' (647; Suisse Romande).
gandille 'shameless woman, streetwalker' (745; Rhône).
gandoio 'streetwalker, shameless woman' (786; Nice, Bayonne).
gandouio 'streetwalker, shameless woman' (786; Nice, Bayonne).
ganela 'woman' (758; Usseglio).

farfaias 'hair-brained young girl' (789; Provençal).
fargognas 'fat, slovenly girl' (789; Provençal).
farramaouco 'maladjusted woman' (921; Tarn).
fastineuse 'wheedler' (57; Wallon).
fatrasse 'meddler' (373; Saintonge).
favioulha 'simple and credulous woman' (647; Suisse Romande).
fawenne 'marten; cunning woman' (57; Wallon).
feme-covert 'married woman' (217; Anglo-Norman).
fèo 'fairy; old woman singular by her attire' (921; Tarn).
feute 'lascivious woman' (57; Wallon).
fignolante 'fashionable, elegant girl' (57; Wallon).
filhandran 'girl of bad life; naughty girl' (921; Tarn).
filhandras 'tall, badly dressed girl' (789; Provençal).
filly 'filly; girl' (217; Anglo-Norman).
fláas 'woman who dresses badly' (183; Gondecourt).
flahutte 'flatterer' (143; Bouillon).
flahutte 'flatterer' (2; Bouillon).
flairante 'stinkard; vain woman who tries to humiliate others' (57; Wallon).
flamik 'rather simple, tall, indolent woman' (204; Saint-Pol).
flamtresse 'woman who chatters in Flemish' (57; Wallon).
flandrine, flandrouille 'slothful, nonchalant woman' (783; Provençal).
flatte 'cow's dung; woman without energy' (57; Wallon).
flauche 'woman who says nothings' (143; Bouillon).
flutresse 'woman who drinks much and often' (57; Wallon).
fondron 'short, fat woman' (453; Côte d'Or).
forcouteie 'woman of the town' (57; Wallon).
forgaou 'untidy girl' (987; Marche).
forpasseie 'woman faded prematurely by libertinage' (57; Wallon).
forsoleie 'woman who wastes money on drinks and feasts' (57; Wallon).
forzarderesse 'spendthrift, bad household manager' (57; Wallon).
fotenne 'toy; unbearable little girl' (57; Wallon).
fouainna 'woman with a curious, malicious, pointed face' (647; Suisse Romande).
fouianna 'curious, indiscreet, pointed-faced woman' (647; Suisse Romande).
fouisso 'fat woman' (789; Provençal).
fourouse 'coquettish woman' (560; Moselle).
foutinette 'prostitute' (26; Paris).
foyon 'mole; loose woman' (57; Wallon).
frachivas 'broad fallow ground; woman who neglects her toilet' (789; Provençal).
franbiassi 'untidy woman' (373; Saintonge).
frawetigneuse 'deceiver, cheater at play' (57; Wallon).
fregasse 'streetwalker' (419; Maine).
fricandella 'live, light young girl' (764; Dauphiné).
fricaudeto 'gay working-girl, coquette, little bourgeoise of bad manners' (786; Nice, Bayonne).
friguenelle 'young woman who seeks delicacies; courtezan' (214; Normandie).
friguette 'elegant, alert girl' (143; Bouillon).
frikette 'light girl, grisette'(57; Wallon).
fringueres 'girls' (756; Loire).
fringuette 'elegant, alert girl' (143; Bouillon).
friperasse-, -osse 'extravagant woman' (560; Moselle).
frise 'mistress, kept woman' (1094; Béarn).

MISCELLANY

drudzon 'girl strong and robust for work' (647; Suisse Romande).
duc-pesey 'mealy-mouthed woman, slow of speech and movement' (204; Saint-Pol).
dulawee 'girl of bad life' (81; Malmédy, Wallon).
dulawèe 'girl of bad life' (57; Wallon).
dzappa 'grumbler' (715; Savoie).
dzappa(na) 'vulgar gossip' (723; Bresse and Bugey).
dzêrêta 'babbling woman' (677; Suisse Romande).
dzeve 'young girl' (758; Jura).
dziga 'tall, alert girl' (675; Vaud).
ebiatze 'girl of bad life' (678; Suisse Romande).
èbiatze 'hussy' (647; Suisse Romande).
ebzinwar 'streetwalker' (204; Saint-Pol).
econneie 'woman of elevated stature' (57; Wallon).
edoirmowe 'asleep; indolent woman' (57; Wallon).
efforcié 'violated woman' (426; Bourgogne).
efouweresse 'sower of discord' (57; Wallon).
egaiouleuse 'coaxer' (57; Wallon).
ejaleie 'cold, unfeeling woman' (57; Wallon).
ekopèère 'tall, thin woman' (182; Gondecourt).
emiceie 'ninny' (57; Wallon).
emprenhade 'animal with young; pregnant woman' (1094; Béarn).
enceinto 'pregnant woman' (828; Provence).
enocaine 'innocent and foolish woman' (57; Wallon).
eplasse 'helpless creature' (57; Wallon).
epufkinneresse 'woman who taints by her filthiness' (57; Wallon).
escamandre 'ragged, shameless woman' (881; Hérault).
escamandre 'devilish, shameless, ragged woman' (789; Provençal).
escrokeuse 'cheat, thief' (57; Wallon).
eserpeno 'woman whose hair is ruffled' (675; Vaud).
estale 'splinter; woman long and thin of stature' (57; Wallon).
ètertineùwe 'woman to whom a lover furnishes money' (151; Wallon-Français).
evairaie 'pregnant woman' (57; Wallon).
évaltonnée 'light girl, tom-boy' (667; Suisse Romande).
evôtrikĕ 'sprite who likes to pursue women; woman of sprightly charm' (541; Meuse, Dombras).
fadrine 'woman of bad life' (1094; Béarn).
fadrine 'woman of bad morals' (1088; Basses-Pyrénées).
faflotte 'pellicle; woman without energy' (57; Wallon).
fafoie 'gossipy, snotty, impertinent girl' (38; Wallon).
fafoie 'gossip' (57; Wallon).
faguenne 'fagot; emaciated woman' (57; Wallon).
faĭasse 'woman who attracts attention by her strange dress and manners' (703; Savoie).
fanfine 'gay working-girl' (57; Wallon).
fangas 'dirty woman' (789; Provençal).
fanguét 'slattern' (783; Provençal).
fanjas 'dirty woman' (805; Dauphiné).
fantômă 'tall, thin woman' (703; Savoie).
farată 'badly dressed wanderer' (703; Savoie).
farbelouse 'poorly dressed woman' (373; Saintonge).
fardau 'badly dressed girl' (789; Provençal).

70 WORDS AND DESCRIPTIVE TERMS FOR 'WOMAN' AND 'GIRL'

documin 'document; old woman who wants to be young again' (57; Wallon).
dōdaine 'fat woman' (677; Suisse Romande).
dodēn 'delicate, tender, sensitive woman' (182; Gondecourt).
dodera 'shallow woman' (712; Savoie).
dolēn(t) 'plaintive, lifeless woman' (182; Gondecourt).
doncel, donzelle 'woman or girl of mediocre state and suspect behavior' (105; Liège).
dondaine 'short, fat, gay girl' (715; Savoie).
dondon 'lively, plump woman' (57; Wallon).
dondon 'short, fat, jovial girl' (745; Rhône).
dondon 'ruddy, plump girl or woman' (151; Wallon-Français).
dondon 'extremely plump woman' (622; Franche-Comté).
dondon 'plump woman' (105; Liège).
dondon 'woman (term of scorn)' (226; Seine Inférieure).
donny 'prostitute' (217; Anglo-Norman).
dorgassi 'insulting name given to the fair sex' (763; Dauphiné).
dorlaînn 'dawdling, nonchalant woman; weeping, sad woman' (105; Liège).
dorlaine 'nonchalant woman, bad housekeeper' (57; Wallon).
dòrn 'girl of bad life' (320; Maine).
doudenne 'foolish, silly woman' (57; Wallon).
doudou 'fiancée, mistress' (57; Wallon).
doudou 'simple girl' (233; le Havre).
douelle 'barrel stave; woman without figure and of bad life' (356; Deux-Sèvres).
doulèro 'woman of bad life' (1023; Gascogne).
doundoun 'fat woman' (813; Drôme).
dourgasso 'nonentity; fat pitcher; insulting term' (813; Drôme).
dourleta 'little woman' (978; Puy-de-Dôme).
dragõ 'dragon; somewhat shrewish woman' (204; Saint-Pol).
dragon 'woman with masculine charm and brusk manners' (57; Wallon).
drapai 'ragged woman' (57; Wallon).
drelho 'slut, sloven; thick mud' (980; Puy-de-Dôme).
drinette 'young girl' (285; Jersey).
drinette 'slip of a girl' (284; Guernesey).
drïngua 'woman of bad morals' (759; Loire).
drinna 'woman' (758; Haute Savoie).
drôdal 'disagreeable old woman' (105; Liège).
drodale 'disagreeable old woman' (57; Wallon).
droie 'woman of bad life' (57; Wallon).
drolesse 'little girl' (356; Deux-Sèvres).
droline 'girl' (414; Berry).
drolière 'little girl' (384; Touraine).
drongade 'dotard' (57; Wallon).
droûdal 'disagreeable old woman' (105; Liège).
drouille 'slut' (164; Wallon).
drouines 'wenches' (226; Seine-Inférieure).
droulasso 'tall girl' (789; Provençal).
droūlieure 'woman of disorderly conduct' (560; Moselle).
droupin 'old woman who is only an obstacle to others' activity' (57; Wallon).
droussain 'sediment; debauchee' (57; Wallon).
droutze 'procuress' (647; Suisse Romande).
drouze 'cross woman' (81; Malmédy, Wallon).
drudjon 'girl strong and robust for work' (647; Suisse Romande).

MISCELLANY 69

curiole 'a girl fond of walking, who is often on the road' (783; Provençal).
curtail 'whore' (217; Anglo-Norman).
cutte 'debauchee' (57; Wallon).
dada 'silly woman' (57; Wallon).
dadaie 'jovial woman' (57; Wallon).
dadaie 'pleasing, foolish, jovial woman' (105; Liège).
dadaine 'silly young girl, innocent girl' (57; Wallon).
dadette 'simple woman' (57; Wallon).
daha 'foolish, giddy woman' (57; Wallon).
dandineûss 'ninny who does not seem clownish' (105; Liège).
dandinress 'ninny who does not seem clownish' (105; Liège).
dàndlotte 'indolent girl, poor girl who lets herself be fondled by all comers' (284; Guernesey).
dandon 'coarse fat woman' (233; le Havre).
decene 'innocent, foolish woman; one who blushes easily' (57; Wallon).
dedoye 'bold, talkative woman' (560; Moselle).
démon 'wicked woman' (57; Wallon).
dernette 'slip of a girl' (284; Normandie, Guernesey).
dèsalêye 'thinly clad girl' (560; Moselle).
desferrée 'girl of loose behavior' (164; Wallon).
desvée 'foolish, extravagant woman' (330; Anjou).
dial 'wicked woman' (57; Wallon).
dialress 'she-devil, shrew' (105; Liège).
dibacheie 'fallen woman' (57; Wallon).
dibachèie 'debauchee, libertine' (105; Liège).
dibacheuse 'woman who bribes, tampers with young girls' (57; Wallon).
diballeuse 'girl who sorts tobacco leaves' (68; Malmédy, Wallon).
dichoeie 'virago, gawky woman' (57; Wallon).
digogei 'giddy woman' (57; Wallon).
digue 'woman of bad life' (214; Normandie).
diguedi 'housekeeper' (284; Guernesey).
dihaineleie 'awkward woman' (57; Wallon).
dihaiowe 'sickly woman' (57; Wallon).
dihireie 'ragged woman' (57; Wallon).
dihoupeie 'dishevelled woman' (57; Wallon).
dikaieie 'woman whose eyes are circled' (57; Wallon).
dilabodeie 'driveling woman' (57; Wallon).
dilahi 'shameless woman' (57; Wallon).
dilofreie 'woman who laments' (57; Wallon).
dīne 'turkey; silly woman; tall woman' (560; Moselle).
dipehi 'squalid woman' (57; Wallon).
disbâcheie 'prostitute' (38; Wallon).
disempanado 'untidy woman' (982; Puy-de-Dôme).
diswaiemeie 'woman whose hair is in disorder' (57; Wallon).
djappa 'scolding woman, tale-bearer' (647; Suisse Romande).
djiebiero 'woman to be scorned' (1002; Limousin, Périgord).
djoume 'ridiculously dressed woman' (667; Suisse-Romande).
doba 'foolish girl or woman' (647; Suisse Romande).
dobèn 'shameless woman or girl' (320; Maine).
dobiche 'old woman' (244; Normandie, Calvados, Bayeux).

coratthira 'hair-brained girl who does nothing but run' (647; Suisse Romande).
corè 'young strumpet' (814; Drôme).
corey 'rather simple, tall, indolent woman' (204; Saint-Pol).
corrasse 'streetwalker, light woman' (560; Moselle).
cosaque 'woman who has masculine charm' (57; Wallon).
coteto 'young girl; little hen' (788; Midi).
cou 'backside; decrepit old woman' (57; Wallon).
coualerino 'woman of bad life' (813; Drôme).
coŭcate 'scandal-monger' (560; Moselle).
coudenas 'skin; woman of bad life' (789; Provençal).
cougliorâou 'mannish girl' (987; Marche).
coumai 'fat, unshapely woman' (57; Wallon).
couotousidou 'woman who serves a lying-in woman' (789; Provençal).
couquihado 'lark; coquettish, alert woman' (789; Provençal).
couquihoun 'shell; woman of small stature' (789; Provençal).
courandière 'streetwalker' (398; Centre).
courettes 'sweethearts' (756; Loire).
couriolo 'streetwalker, girl of little modesty' (786; Nice, Bayonne).
courlis 'woman who is always out' (353; Vendée, Ile-d'Elle).
courpatasso 'skinny old ewe; ugly old woman' (789; Provençal).
couruso 'streetwalker, girl of little modesty' (786; Nice, Bayonne).
couteto 'pullet; young girl' (789; Provençal).
coutolo 'woman of the common people who affects the fine lady' (789; Provençal).
coutrolo 'woman whom one trusts thoughtlessly' (828; Provençal).
covert 'married woman' (217; Anglo-Norman).
cowe 'she-devil' (57; Wallon).
craipette 'wretched little girl' (57; Wallon).
crale 'disgusting woman' (57; Wallon).
crale 'disgusting, hideous woman' (81; Malmédy, Wallon).
crapaute 'young girl, mistress, (57; Wallon).
crape 'ulcer; debauchee, bad housekeeper' (57; Wallon).
crapo 'girl whose younger sister is married' (805; Dauphiné).
craquette 'liar' (57; Wallon).
crawe 'hook, butt-end of a gun; deformed woman' (57; Wallon).
créature 'woman, servant' (214; Normandie).
créature 'term of insult applied to a girl' (330; Anjou).
créature 'woman' (376; Franche-Comté).
créatures 'women' (349; Vendée).
criquet 'little woman' (57; Wallon).
crôle 'old woman' (356; Deux-Sèvres).
crone 'old woman' (217; Anglo-Norman).
crope-es-cindes 'kitchen drudge; dirty woman' (57; Wallon).
cuck-quean 'whore' (217; Anglo-Norman).
cueisse 'devout girl' (164; Wallon).
cuque 'woman hermit who lives like a savage' (1098; Béarn).
cuque 'mill-moth; woman hermit who lives like a savage' (1094; Béarn).
cur 'squalid woman, one without principles' (57; Wallon).
curai 'squalid woman, one without principles' (57; Wallon).
cureie 'dead beast; woman who no longer has any shame' (57; Wallon).
curie 'thoroughly dirty woman, slut' (164; Wallon).

chinne 'detestable woman; dog' (57; Wallon).
chino 'bitch; avaricious or shameless woman' (789; Provençal).
chipie 'avaricious woman' (445; Saone-et-Loire).
chipie 'shrew' (214; Normandie).
chipote 'haggler' (57; Wallon).
chiquette 'puny slip of a girl' (284; Guernesey).
chnapeuse 'heavy woman drinker of liquors' (57; Wallon).
chorleuse 'woman who courts men' (57; Wallon).
chornia 'woman shamefully dissolute' (858; Basses-Alpes).
chorniassa 'woman shamefully dissolute' (858; Basses-Alpes).
chouftresse 'coaxing woman' (57; Wallon).
choukteuse 'woman who hoots' (57; Wallon).
choulate 'sad woman' (57; Wallon).
chouma 'old she-ass, injurious epithet when addressed to a woman' (647; Suisse Romande).
ciouma 'old she-ass, injurious epithet when addressed to a woman' (647; Suisse Romande).
cipi 'shrew' (182; Gondecourt).
citrouille 'slut, woman of bad character' (667; Suisse Romande).
clabot 'little bell; grumbler' (57; Wallon).
clapette 'gossip, slanderer' (57; Wallon).
clapoteuse 'gossip' (57; Wallon).
claque 'indolent, dirty, lazy woman' (176; Lille).
claque 'woman who dresses with much negligence' (164; Wallon).
cliche d'ouhe 'door button; debauchee' (57; Wallon).
clime 'nonchalant woman' (2; Bouillon).
clime 'woman without vigor' (143; Bouillon).
climena 'streetwalker, concubine' (647; Suisse Romande).
cobanne 'simple, naive, rather stupid woman' (300; Bretagne, Rennes).
còcate 'chatterer' (560; Moselle).
cochêne 'woman familiar in a house' (164; Wallon).
cocota 'cocotte, hen; light woman' (759; Loire).
cocote 'light woman' (151; Wallon-Français).
cocotte 'courtezan' (26; Paris).
coiene 'rind; ill-formed woman' (57; Wallon).
coieteuse 'woman who swears habitually' (57; Wallon).
côka, côke 'ridiculous and boring old gossip' (647; Suisse Romande).
côke 'ridiculous and boring old gossip' (647; Suisse Romande).
cokesante 'gay, lively woman' (57; Wallon).
cokkahssa 'ridiculous giggler' (647; Suisse Romande).
colowe 'adder; sarcastic woman who answers all insults' (57; Wallon).
colso 'streetwalker' (789; Provençal).
connelle 'young lady, madam' (758; Lorraine).
côouseto 'chit-chat' (987; Marche).
copète 'chatterer' (560; Moselle).
copineuse 'one who loves to chat' (57; Wallon).
copurneuse 'debauched woman' (57; Wallon).
côque 'old woman' (703; Savoie).
coquette 'coquette, woman who seeks to please' (783; Provençal).
coquine 'rascal' (198; Boulonnais).
corathire 'madcap girl' (715; Savoie).

66 WORDS AND DESCRIPTIVE TERMS FOR 'WOMAN' AND 'GIRL'

chacaras 'woman lacking good sense' (828; Provence).
chacaras 'woman who has neither bearing nor sense' (788; Languedoc, Provence).
chacha 'fool, babbler' (57; Wallon).
chachalone 'indolent woman' (57; Wallon).
chacharas 'woman without bearing or sense' (789; Provençal).
chachoule 'woman who weeps' (57; Wallon).
chafooutre 'girl who has bad carriage' (858; Basses-Alpes).
chaffette 'babbler' (552; Lorraine).
chaftiress, chaftress 'frivolous and gossipy woman' (38; Wallon).
chaipiowe 'small sickly woman' (57; Wallon).
challée 'tall, unkept woman' (486; Champagne, Aube).
chamarett 'woman who chatters' (38; Wallon).
chamarette 'babbler, indiscreet grisette of brilliant dress' (57; Wallon).
chamiau 'camel; tall, ugly, ill-formed woman' (57; Wallon).
chamô 'slattern; a masculine, dirty, ugly, old, and brutal woman' (105; Liège).
chamous 'streetwalker' (858; Basses-Alpes).
chamousel 'streetwalker' (788; Midi).
chanbranlŏ 'tall, thin woman' (703; Savoie).
chantresse 'gossip' (57; Wallon).
çharcŏ 'woman of questionable morals' (703; Savoie).
charée 'woman of bad life' (214; Normandie).
charogne 'hag; scornful, shameless woman' (57; Wallon).
charospo 'prostitute' (786; Nice, Bayonne).
charoupia 'woman of bad life' (858; Basses-Alpes).
çharvò 'woman of questionable morals' (703; Savoie).
chatarasso 'ugly or bad woman' (786; Nice, Bayonne).
chatasso 'tall girl, fat girl' (789; Provençal).
chato 'young girl' (838; Vaucluse).
chato 'young girl' (788; Provence).
chatouneto 'little girl' (838; Vaucluse).
chatouneto 'little girl' (789; Provençal).
chatouno 'little girl' (838; Vaucluse).
chaupinado 'prostitute' (789; Provençal).
chausson 'woman of bad life' (398; Centre).
chautrinasso 'abandoned woman' (789; Provençal).
chautrine 'slut' (783; Provençal).
chavan 'woman of questionable morals' (703; Savoie).
chawate 'screech-owl, owl; ruffle-haired woman' (560; Moselle).
chawi 'ugly, badly built woman' (57; Wallon).
chena 'basket; debauchee' (57; Wallon).
chendrillon 'dirty woman' (233; le Havre).
chérpinnă 'bargaining woman' (703; Savoie).
chérvŏ 'woman of questionable morals' (703; Savoie).
chicâd 'sniveler, whimperer' (38; Wallon).
chicaneûss 'quibbler; one who loves wrangling' (105; Liège).
chicanress 'quibbler; one who loves wrangling' (105; Liège).
chicate 'woman who weeps' (57; Wallon).
chiffon 'manhandled woman' (57; Wallon).
chifode 'bungler' (57; Wallon).
chiftowe 'chubby woman, one whose cheeks are chubby or swollen' (57; Wallon).
chinchoun 'thin, sickly young girl' (789; Provençal).

cancorna 'grumbler' (756; Loire).
cancouine 'old woman, old skeleton' (667; Suisse Romande).
cancuare 'giddy young girl' (622; Franche-Comté).
cande 'prattler' (136; Saint-Hubert).
canlette 'talker, relater' (164; Wallon).
cannadozeuse 'cajoling woman' (57; Wallon).
canoie 'lazy woman' (57; Wallon).
canou 'old woman' (57; Wallon).
cantino 'canteen; bottle case; old woman' (789; Provençal).
cantonière 'woman of bad life' (783; Provençal).
caqueteuse 'woman who chatters' (38; Wallon).
carase 'woman of the torn' (57; Wallon).
caraud 'bad plot; old woman' (356; Deux-Sèvres, Bas-Gâtinais).
carcagno 'grumbler' (828; Provence).
carcano 'old woman' (788; Provence).
carcasse 'thin woman' (57; Wallon).
carogne 'woman of bad life' (745; Rhône).
carogno 'corpse; debauchee' (789; Provençal).
carougnas 'public girl' (789; Provençal).
carounhe 'vile woman' (1088; Basses-Pyrénées).
carpigno 'quarrelsome woman' (789; Provençal).
carqueto 'old ewe; mild old woman, toothless old woman' (789; Provençal).
carrello 'wheelbarrow;' *misè carello* 'grumbler' (789; Provençal).
carroigne 'vile woman' (1088; Basses-Pyrénées).
cascarlet 'frivolous, giddy woman' (789; Provençal).
caspouieusse 'woman who squanders her wealth' (57; Wallon).
cassemousseuse 'intriguing woman' (57; Wallon).
cassot 'needle box; paper bag; old cow; old woman' (233; le Havre).
catacraise 'old woman' (330; Anjou).
catan 'woman of ill repute' (214; Normandie).
catan 'woman of bad life' (330; Anjou).
catarino 'prostitute' (786; Limousin).
catasso 'cat; double-faced woman' (789; Provençal).
catau 'girl of bad life; doll' (448; Bourgogne).
catharreuse 'woman disposed to debauchery' (330; Anjou).
catienne 'chatterer' (662; Franche-Comté).
catiule 'sickly, puny, complaining woman' (703; Savoie).
catot 'woman who is nothing' (353; Vendée).
catou 'doll; girl of loose living' (176; Lille).
cattimaula 'boring, repeating, sad, complaining woman' (647; Suisse Romande).
caulêne 'idle, lazy slanderer' (553; Belgique).
cavala 'mare; insulting term applied to women' (759; Loire).
cavalasso 'abandoned woman' (789; Provençal).
cavale 'mare; prostitute; tall woman' (57; Wallon).
câye 'idle, frivolous girl or woman' (552; Lorraine).
cayriè 'towel; bucking-cloth; dirty woman' (921; Tarn).
cayrièras 'women of bad life' (921; Tarn).
chabraca 'woman lacking poise' (858; Basses-Alpes).
chabraco 'tall, ungainly woman' (980; Puy-de-Dôme).
chabrak 'shameless woman, prostitute' (105; Liège).
chabraque 'streetwalker' (57; Wallon).

brocson 'woman of coarse manners, one who dresses in bad taste' (214; Normandie).
brodion 'meddler' (56; Wallon).
brognate 'sulker' (57; Wallon).
brott 'dissolute woman' (105; Liège).
brotte 'bitch in heat; mannerless, immoral woman' (57; Wallon).
brouffetresse 'woman who loves a good time' (57; Wallon).
brouhagne 'barren cow; sterile woman' (57; Wallon).
broukeie 'old sheep; decrepit old woman' (57; Wallon).
broumo 'phlegm; woman of bad life' (789; Provençal).
brubette 'mistress, lover' (57; Wallon).
brujo 'jar; fat woman' (789; Provençal).
bru-mâle 'very fat woman who is childless' (384; Touraine).
bruye 'scorned woman' (678; Suisse Romande).
bubrasse 'little girl who likes to play only with boys' (622; Franche-Comté).
bulo, bulot 'fat, greasy person' (204; Saint-Pol).
burdōdìn 'streetwalker' (204; Saint-Pol).
burgàdin 'streetwalker' (204; Saint-Pol).
burtress 'woman who drinks habitually' (57; Wallon).
buze 'woman without intellect' (57; Wallon).
bzinwar 'streetwalker' (204; Saint-Pol).
cabasse 'term of insult when applied to a woman' (330; Anjou).
cabau 'capital; light young girls' (789; Provinçal).
cabestre 'halter; shameless girl' (858; Basses-Alpes).
cabrak 'tall, ill-formed woman of low intellect' (204; Saint-Pol).
cabrasso 'big goat; fickle, dissipated girl' (789; Provençal).
caburier 'woman in disorder' (182; Gondecourt).
cacarette 'chatterer, prude; streetwalker' (57; Wallon).
cache 'sow that has undergone the operation of ablation of the ovaries; rather selfish woman' (560; Moselle).
cadelasso 'tall young girl who likes a good time' (789; Provençal).
cafougneuse 'one who rumples all that she touches' (57; Wallon).
caftiresse 'woman who loves coffee excessively' (57; Wallon).
cagneteuse 'quarrelsome woman' (57; Wallon).
cagniesse 'sensitive, easily-offended woman' (57; Wallon).
cahute 'prostitute' (57; Wallon).
caillette 'prattler' (330; Anjou).
cakaïe 'debauchee, scum of the earth' (105; Liège).
cakante 'eager person, girl who takes care of herself' (57; Wallon).
cakète 'tattler' (57; Wallon).
cakteûss 'chatterer, babbler, slanderer' (105; Liège).
caktress 'chatterer, babbler, slanderer' (105; Liège).
calaude 'prattler' (136; Saint-Hubert).
cale 'bonnet, former headdress of old women; girl of low condition' (448; Bourgogne).
calèssio 'tiresome, troublesome woman; parasite' (921; Tarn).
calignairis 'marriageable young girl' (789; Provençal).
camache 'old objects; old woman in rags' (57; Wallon).
camanette 'tattler, scandal-monger' (176; Lille).
cameou 'camel; woman of bad life' (858; Basses-Alpes).
camoche 'flat-nosed woman' (703; Savoie).
camou 'camel; woman of bad life' (204; Saint-Pol).
campiï 'woman who governs her family with authority' (758; Savoie).

bougonneuse "grumbler' (299; Bretagne, Rennes).
bougresse 'sad woman' (760; Loire).
bougresse 'hussy' (57; Wallon).
bouhon 'bush; foolish, stupid woman' (57; Wallon).
bouimenco 'woman dishonest in business' (828; Provence).
bouiro 'woman in child-bed' (789; Provençal).
boulliaco 'slovenly woman' (991; Limousin, Périgord).
bouloufe 'fat woman' (57; Wallon).
bourache 'drunken woman' (57; Wallon).
bourdresse 'liar' (57; Wallon).
bourgatta 'annoying woman who pries' (647; Suisse-Romande).
bourgeoise 'woman, wife' (142; Ardenne).
bourgeoise 'lady of the house, wife' (384; Touraine).
bourik 'she-ass; very stupid person' (38; Wallon).
bourique 'ass; foolish woman' (57; Wallon).
bouriquote 'she-ass; stupid or ignorant woman' (448; Bourgogne).
bourotte 'little ruddy-faced woman' (57; Wallon).
bourreaude 'cruel woman or girl' (667; Suisse-Romande).
bouta 'impudent little girl' (57; Wallon).
bovardo 'chatterer' (987; Marche).
bovouso 'envious person' (987; Marche).
boyaude 'girl, child' (373; Saintonge).
boye 'girl, child' (373; Saintonge).
bracă 'scatterbrained, giddy girl' (703; Savoie).
bradye,-er 'squanderer' (204; Saint-Pol).
brafă 'vivacious, talkative, unreflecting woman' (703; Savoie).
bràkà 'bizarre, whimsical woman' (703; Savoie).
brakleuse 'one who habitually boasts, shouts' (57; Wallon).
brasqua 'woman' (807; Champsaur).
brasqua 'woman' (758; Dauphiné).
brasse 'mother who carries her child in or on her arms' (1094; Béarn).
bredas 'busybody' (356; Deux Sèvres).
bredasse 'busybody' (356; Deux Sèvres).
bredasse 'careless woman' (353; Vendée, Ile-d'Elle).
bredel 'bad housekeeper' (320; Maine).
bredouille 'stammerer; careless woman who acts without thinking' (703; Savoie).
bregotsire 'woman slovenly in her work' (675; Vaud).
brèheugne 'sterile woman' (560; Moselle).
brèhhelate, -ote 'young girl' (560; Moselle).
breiatte 'brawler, wrangler' (57; Nallon).
brekola 'prattler' (715; Savoie).
brěkola 'prattler' (677; Suisse Romande).
breleque 'little girl' (used only in a bad sense) (233; le Havre).
brelinganne 'girl without good carriage' (419; Maine).
breloka 'unreliable, talkative woman' (647; Suisse Romande).
brětă 'woman who speaks without thinking' (703; Savoie).
breza 'jade' (622; Franche-Comté).
bribresse 'beggar, lazy woman' (57; Wallon).
bride 'fiancée, bride' (217; Anglo-Norman).
briquotte 'woman whose clothes are patched and pieced' (622; Franche-Comté).
briyulà, awkward, unreflecting woman' (703; Savoie).

bisa 'north wind; black, swarthy, emaciated woman' (647; Suisse-Romande).
bisteu 'any beast with horns; clumsy woman' (57; Wallon).
bitou(no) 'pig; abandoned girl' (789; Provençal).
blaboñèi 'woman' (182; Gondecourt).
blagæïṽ, blagæ̂z 'gossip' (204; Saint-Pol).
blamante 'flaming; girl full of life and ardor' (57; Wallon).
blazineuse 'streetwalker' (57; Wallon).
bléfaute 'driveler' (151; Wallon-Français).
blĭôsë 'lazy woman who loves to be pampered' (703; Savoie).
bliwette 'spark; impudent, talkative little girl' (57; Wallon).
blokai 'short, very fat woman' (105; Liège).
blonde 'amorous mistress' (57; Wallon).
blonde 'sweetheart; fiancée' (560; Moselle).
blonde 'sweetheart, mistress' (448; Bourgogne).
bloquai 'block; little fat woman' (57; Wallon).
bobanne 'simple, naive woman' (300; Bretagne, Rennes).
bobos 'hunchback' (204; Saint-Pol).
bobòt 'narrow-minded woman, particularly in religion' (320; Maine).
bocque 'little woman' (342; Poitou).
bodale 'short, very fat woman' (151; Wallon-Français).
bodeie 'short, massive woman' (57; Wallon).
boetsa 'woman' (758; Haute Savoie).
bofièsse 'fat, chubby-cheeked woman' (560; Moselle).
boice 'married woman' (758; Jura).
boilla 'young girl' (742; Lyon).
boio 'young girl' (789; Provençal).
bokajer 'girl courted in view of marriage' (320; Maine).
bolin bolia-ou 'careless woman' (991; Limousin, Périgord).
bondrée 'woman short and fat as a sluice' (214; Normandie).
bonĭĕ 'soft, indolent girl' (703; Savoie).
bonne 'sweetheart' (667; Suisse-Romande).
bonniche 'maid-of-all-work' (667; Suisse-Romande).
boque 'fat, chubby-cheeked girl' (748; Lyon).
borbotâna 'woman who never ceases murmuring' (675; Vaud).
borghatta 'annoying woman who pries' (647; Suisse-Romande).
borirë 'churn; small, slender woman' (703; Savoie).
borlina 'courtezan' (758; Alpes Piémontaises).
borlio 'one-eyed woman' (871; Gard).
borotă 'small, slender woman' (703; Savoie).
bouama 'bad woman' (808; Hautes-Alpes).
bouassa 'woman' (758; Savoie).
bouba 'little girl' (647; Suisse-Romande).
boubetta 'very little girl' (647; Suisse-Romande).
boubou 'brush; woman whose hair is unkept' (57; Wallon).
boucanneresse 'quarrelsome woman' (57; Wallon).
boudion 'baby; woman of very small stature' (57; Wallon).
bouéba 'little girl' (647; Swiss-Romande).
bouginjouno 'rather fat little woman' (813; Drôme).
bouelle 'young girl' (393; Loiret).
bougnassou 'woman, girl who squats by the fire' (921; Tarn).
bougonneuse 'constant scolder' (632; Jura).

MISCELLANY

bavette 'one who talks too much, saying little' (233; le Havre).
bazaruetto 'meddler; liar' (828; Provence).
bazou 'woman without order or charm' (171; Marche-lez-Ecaussines).
baztotu 'good-for-nothing woman' (309; Basse-Bretagne).
beato 'woman who leads an exemplary life' (828; Provence).
becudo 'impertinent, gossiping woman' (1023; Gascogne).
bédèle 'joker, prattler' (579; Vosges).
bèdèle 'prattler' (532; Razey).
bedodene 'very fat woman' (675; Vaud).
bedouma 'simple, unskilled, lazy woman' (647; Suisse-Romande).
bedoume 'simpleton' (667; Suisse-Romande).
beduma 'fickle young girl' (686; Suisse-Romande).
beduma 'woman of low intelligence' (675; Vaud).
bèeusse 'little girl' (532; Pexonne).
bega 'courtezan' (758; Alpes Piémontaises).
begnule 'foolish, awkward, languid woman' (703; Savoie).
begœl 'gossip, slanderer' (204; Saint-Pol).
begueinn 'affectedly devout woman; nun of the Pays-Bas, principally of Liège; devout, pious, superstitious woman' (38; Wallon).
béguene 'devout woman' (57; Wallon).
bekasin 'simple, talkative woman' (204; Saint-Pol).
belo 'girl or woman' (917; Toulouse).
benĭulă 'awkward, foolish, languid woman' (703; Savoie).
bĕōslier 'woman badly dressed up' (182; Gondecourt).
bèque 'sow; insulting term for woman' (667; Suisse-Romande).
berbī, barbī 'sheep; easily-led woman lacking in sense and energy' (204; Saint-Pol).
berbī, berlod 'woman lacking will and energy' (204; Saint-Pol).
berdale 'woman of irregular conduct' (214; Normandie).
berdasso 'foolish babbler' (384; Touraine).
berdolle 'prattler' (622; Franch-Comté).
berdoule 'woman without order' (176; Lille).
berken 'woman without will or energy' (204; Saint-Pol).
berlingue 'babbler, indiscreet talker' (164; Wallon).
bernada, barnada 'old woman who throws grain on the bride' (632; Suisse-Romande).
berzole 'giddy woman' (214; Normandie).
bĕtelate 'little animal; stupid woman' (560; Moselle).
beurdinne 'woman who talks fast and much' (560; Moselle).
beuresse 'woman who drinks much and often' (57; Wallon).
beyar, bewar 'woman who looks simple and whose mouth hangs open, one with indiscreet curiosity' (204; Saint-Pol).
bèyerasse, bèyeresse 'one who loves to give' (560; Moselle).
bezotsire 'woman who ruins her household' (675; Vaud).
bianche-tête 'female, especially old woman' (560; Moselle).
bidou 'sweetheart, girl to whom one pays court' (57; Wallon).
biebo 'unskilled, ignorant woman' (275; Normandie, Manche).
biêla 'silly woman' (759; Loire).
biesse 'foolish woman' (57; Wallon).
bigeot, bigeotte 'gay young woman' (622; Franche-Comté).
bihe 'hind; ungainly woman; streetwalker' (57; Wallon).
biko 'goat, hind; insulting term when applied to a woman' (1023; Gascogne).
biotsa 'satirical, mocking, curious young girl' (647; Suisse-Romande).

baban 'slow, tiresome woman' (320; Maine).
babāye 'silly, foolish woman' (560; Moselle).
babette 'babbler' (522; Champagne, Ardennes).
bâbi 'blockhead' (667; Suisse-Romande).
babiêla 'silly woman' (759; Loire).
babille 'chatterer' (448; Bourgogne).
babinette 'gossip' (57; Wallon).
babiota 'little girl' (858; Basses-Alpes).
baboilla 'ignorant and foolish babbler' (394; Loiret).
badarela 'woman who cries out often' (858; Basses-Alpes).
badlagoule 'talker' (284; Guernesey).
badoire 'talkative woman' (398; Centre).
badou 'stone cask; fat woman' (57; Wallon).
badouere 'talkative woman' (398; Centre).
badoulette 'healthy, plump woman' (176; Lille).
badrey 'tall, somewhat simple woman' (204; Saint-Pol).
bâicûss 'idler, but one who admires all with a frivolous curiosity; fool, silly' (105; Liège).
baile 'nurse of a young girl' (763; Dauphiné).
bâiress 'idler, silly; fool' (105; Liège).
bajawe 'gossip' (57; Wallon).
balarme 'giantess' (57; Wallon).
baléque 'gossip' (214; Normandie).
balzineûss 'loafer' (105; Liège).
bambane 'soft, apathetic woman' (373; Saintonge).
banchée 'young girl of whom the bans have been published' (397; Berry).
bandrol 'woman of light morals' (204; Saint-Pol).
bano 'woman to whom a cavalier gives his arm' (980; Puy-de-Dome).
banstai 'girl who has lost her honor' (57; Wallon).
barada 'rustic cap or bonnet; thoughtless, giddy young girl' (57; Wallon).
barada 'giddy, foolish, inconsiderate woman' (105; Liège).
barbeintsa 'old woman' (758; Savoie).
barbella 'talkative woman' (756; Loire).
barbott 'scolding old woman' (105; Liège).
bardahe 'pole, rod; restless, fidgety woman' (57; Wallon).
bardjaka 'indiscreet, garrulous woman' (647; Suisse-Romande).
bardotsé 'woman who does her work carelessly' (675; Vaud).
baritella 'foolish young girl' (373; Saintonge).
barjagua 'talkative woman' (759; Loire).
barjaka 'indiscreet, garrulous woman' (647; Suisse-Romande).
bartavella 'prattler, thrush, red partridge' (759; Loire).
baruçhĕ 'tall, thin woman' (703; Savoie).
bas-cou 'little one' (57; Wallon).
baset 'woman of small stature' (204; Saint-Pol).
bastoun-vesti 'very thin woman' (858; Basses-Alpes).
batan-pèno 'great talker' (789; Provençal).
baudrelo 'prostitute, flabby woman' (786; Nice, Bayonne).
baurbette 'bearded woman' (57; Wallon).
bautse 'woman' (758; Suisse-Romande).
bavardress 'prattling busybody' (105; Liège).
bavaresses 'talkers' (226; Seine-Inférieure).

Chapter XXVII

MISCELLANY

abandounado 'abandoned woman' (1023; Gascogne).
abbesse 'woman who keeps a house of ill fame' (57; Wallon).
adawiante 'fascinating woman' (57; Wallon).
adawieuse 'coaxing woman' (57; Wallon).
afiloteûss 'swindler, pickpocket, thief' (105; Liège).
afilouteuse 'cheat, swindler' (57; Wallon).
afiloutress 'swindler, pickpocket, thief' (105; Liège).
afrontaie 'bold, impudent, insolent woman' (105; Liège).
afronteûss 'cheat, deceiver' (105; Liège).
agnes 'idiot' (57; Wallon).
agnèsse, agneusse 'woman of little judgment' (560; Moselle).
ahess 'courtezan, streetwalker' (105; Liège).
ahesse 'utility; courtezan' (57; Wallon).
aidketeuse 'woman who makes marriages' (57; Wallon).
aloie 'coquette, flirt' (57; Wallon).
amazônn 'amazon, warrior woman' (38; Wallon).
ameresse 'sweetheart' (104; Liège).
amour 'love, sweetheart' (104; Liège).
anbaiteûss 'tiresome person, coaxer, deceiver, wheedler' (105; Liège).
anchelle 'young lady' (104; Liège).
anguigne 'resourceless woman, somewhat idiotic' (448; Bourgogne).
anièsse 'she-ass; woman of little intelligence' (560; Moselle).
antion 'heavy, awkward woman' (453; Côte d'Or).
anturlûr 'wench, woman or girl to be scorned' (105; Liège).
ănūs 'worn-out old woman' (204; Saint-Pol).
anweie 'eel; lively young girl who plays with men' (57; Wallon).
aplaquante 'irritating woman, one who seeks marriage' (57; Wallon).
argine 'ugly, avaricious woman' (57; Wallon).
argoteie 'very sly woman' (57; Wallon).
ârgoteûss 'quibbler' (105; Liège).
ârgotress 'quibbler' (105; Liège).
armuenè 'almanach; talkative woman' (622; Franche-Comté).
arpalhan 'tall, thin woman' (858; Basses-Alpes).
arpiano 'woman remarkable for her claws; thief' (789; Provençal).
ărply 'bad and loud woman' (204; Saint-Pol).
arrehillo 'granddaughter' (788; Midi).
artifae 'old gadget; hideous old woman' (57; Wallon).
augatte 'foolish girl' (532; Landremont).
aurimiel 'oriole; fine and accessible woman' (57; Wallon).
avaricûss 'avaricious, stingy, tenacious woman' (105; Liège).
awaton 'courtesan who has had a child by a married man' (38; Wallon).
awatron 'one who neglects her household' (57; Wallon).
awe 'goose; foolish, ignorant woman' (57; Wallon).
awi 'woman easily led' (204; Saint-Pol).
awihon 'sting; irritating woman' (57; Wallon).

ones are at numbers 53, 63, 64, 66, 67, 188, 208, 976, 977, in Doubs, Vosges, Ardennes, and Switzerland.

ALF 1226 'servante' includes a few of interest, as: *bās* (Manche) at 386, *bāsēl* (Meurthe-et-M.) at 181.[163]

[163] *A* and *e* of the second term each bears a grave accent directly above the long ark.

TERMS DERIVED FROM *BACASSA, *BAGASSA, BACHES 57

594,[134] *ba:slat* (bāslat) 594,[135] *bajasse* 190, *basse* 282, *baisse* 284, *basselette* 186, *baiasse* 478, *beyes* 558,[136] *bweyes* 579,[137] *pëieusse* 583, *beyes* 584,[138] *bacelle* 143, *bouaichelle* 143, *bäsel* 572,[139] *baicelle* 583, *bâcelle* 583, *basselette* 186, *bachelette* 175, *bouaichelette* 143, *basselette* 186, *bachelette* 175, *bāslot* 574,[140] *baslot* 579,[141] *béçote* 614, *baissote* 620, *besot* 624,[142] *besot* 625,[143] *baisstot* 627, *bäsat* 617,[144] *bäseta* 669,[145] *bachelése* 21, *bagasso* 916, *bās*,[146] *bacelle*,[147] *bauchelle*,[148] *bas* 'vieille fille',[149] *bēs*,[150] *beyes*,[151] *bēs*,[152] *bâcèle*,[153] *bosel*,[154] *b(r)essayla*,[155] *brexlat*,[156] *bacelote*,[157] *bāslot*,[158] *bāsot*,[159] *béçate*,[160] *besot*,[161] *baista* 'fille grande et vigoureuse'.[162]

ALF 570 'fille' includes some scattered terms of which a few interesting

[134] Plombières (Vosges). In the source *a* resembles the mathematical symbol for infinity with line ends shorter in length and the enclosure made more in the shape of an italicized *a*; *o* is printed as Fr. phonetic symbol for open *o* (i.e., like *c* opening to the left); within the parentheses *o* is as in note 106.
[135] Les Gragnes (Vosges). Initial *a* as in note 134.
[136] Each *e* as in note 61.
[137] See note 136.
[138] See note 136. Belmont (Bas-Rhin).
[139] *E* as in note 61.
[140] *O* as in note 106.
[141] *O* as in note 106.
[142] *E* bears a dot directly under it; *o* as in note 106.
[143] See note 142.
[144] *S* bears a small v-shaped symbol directly over it.
[145] *E* as in note 142.
[146] Bessin (district in Calvados), FEW 1.196–7.
[147] Picard, FEW loc. cit.
[148] Carignan (Ardennes). It is suggested that Mid. Fr. *vasselle*, Mod. Fr. *valet*, Champ. *vacelle* have influenced such forms as *bacelle*, etc., FEW loc. cit.
[149] Malmédy (Belgium), FEW loc. cit.
[150] La Baroche (Haut-Rhin); in addition to the regular meaning of 'girl' this term also means 'daughter'. Initial *e* as in note 61.
[151] La Baroche (Haut-Rhin). Each *e* as in note 61.
[152] Orbey (Haut-Rhin). *E* as in note 61.
[153] Wallon (Belgium).
[154] Namur (Belgium). *O* bears a dot directly under it; *s* bears a tiny hook directly under it opening toward the top and also bears a v-shaped symbol of the same size directly above it.
[155] Switzerland.
[156] Metz (Moselle). *E* as in note 61; in the source *x* is printed in the form of a capital *x* in italics.
[157] Metz (Moselle).
[158] *O* as in note 106. Southern Vosges.
[159] Vosges and Southern Vosges. *E* as in note 61.
[160] Doubs.
[161] Sancey (Doubs). *E* bears a dot directly under it; *o* as in note 106.
[162] Jura Bernois.

bācot,[107] bācot,[108] bācot,[109] bēcnot,[110] bacot,[111] bacnot,[112] bécat,[113] bècnat,[114] bèsat,[115] bècnat,[116] basèt,[117] bäsét,[118] bésèta,[119] bésta,[120] bèchat',[121] baichotta,[122] baichetta,[123] baisotte,[124] bassoutotte,[125] bésot,[126] baiss'tot,[127] bessaula,[128] bechaula,[129] bressaula,[130] besāoula,[131] brechālā 'jeune fille nubile',[132] besoleta[133] ba:slot (bāslot)

[107] Ibid., Joncherey (Belfort). *C* as in note 90.

[108] Ibid., Jura Bernois at ALF 74. Pauli, pp. 151–2, refers to Lorraine forms *gacot* and *gacnot* (*c* in each as in note 90), which words he cites as used in varying forms to mean 'jeune fille' in the dépts. of Meuse, Meurthe-et-Moselle. In Haute-Marne, Vosges, Haute-Saône, Côte d'Or, Doubs, Evêché de Bâle the forms *gacot, gacnot, gacoet, gacot, gasoet, gacel* (*c* in each of these as in note 90) are used to signify 'jeune fille'. These last forms, observes Pauli, are used in places where *garce* is used in a pejorative sense. *C* as in note 90.

[109] Ibid., Jura Bernois at ALF 73. *Fēy* is sometimes with the same sense; generally in this region: *bāsnot* (*s* with v-shaped symbol directly over it and *n* with small arc directly under it opening toward the right), possibly through influence of East. Fr. *gachenet* 'garçonnet', FEW 1.197. *C* as in note 90.

[110] Ibid., Jura Bernois at ALF 73. *C* as in note 90.

[111] Ibid., Jura Bernois at ALF 72. *C* as in note 90; *a* bears a normal sized printed *e* directly over it.

[112] Ibid., Jura Bernois at ALF 72. *C* as in note 90; *a* bears a normal sized printed *e* directly over it.

[113] Ibid., Jura Bernois at ALF 71. *C* as in note 90.

[114] Ibid., Jura Bernois at ALF 71. *C* as in note 90.

[115] Ibid., Jura Bernois at ALF 64. *A* bears a tiny circle directly above it.

[116] Ibid., Jura Bernois at ALF 64. *C* as in note 90; *a* as in note 115.

[117] Ibid., Neuchâtelois at ALF 63.

[118] Ibid., Neuchâtelois as cited by Meyer-Lübke.

[119] Ibid., Jura Bernois, region of canton on Neuchâtel bordering on canton of Vaud, and sporadically in canton of Fribourg.

[120] Ibid., see note 119.

[121] Ibid., see note 119.

[122] Ibid., Jura.

[123] Ibid., see note 122.

[124] Ibid., in neighboring parts of Doubs and in Montbéliard.

[125] Ibid., see note 124. Pauli quotes Contejean on the fact that this word is used only at Montagne.

[126] Ibid., Damprichard.

[127] Ibid., Fourgs.

[128] Ibid., Fribourgeois. Pauli remarks that all Suisse Romande patois have a derivative in *-ola*. He also cites Gauchat's observation that there is a word *bresāoula* (in source *e* is printed as Fr. mute *e* symbol and *ou* are slightly raised) that means a part of a carriage [braceola], which may be responsible for the intrusive *r*.

[129] Ibid., see note 128.

[130] Ibid., see note 128.

[131] Ibid., canton of Vaud, surviving today only in the canton of Fribourg. *E* is printed as symbol for Fr. mute *e; ou* are slightly raised.

[132] Ibid., Pauli cites this as the pronunciation of *besāoula* (in source, *e* symbol for Fr. mute *e* and *ou* slightly raised) at Guyère. *E* is printed as symbol for Fr. mute *e*.

[133] Ibid., see note 132. Initial *e* is printed as symbol for Fr. mute *e*.

with the attractions of a woman, hermaphrodite', *bachelo* 788 'bachelette, jeune fille, servante',[76] *bagasse* 789 'woman who leads an evil life', *boglies*,[77] *bèyès*,[78] *bwèyès*,[79] *bas*,[80] *bèyès*,[81] *béeusse*,[82] *beyes*,[83] *bacelle*,[84] *bacèl*,[85] *basèl*,[86] *bachelo* 'servante, jeune fille',[87] *basselle (bāsel)* 'jeune fille',[88] *bauchelle*,[89] *bwecel*,[90] *bwêsal*,[91] *bwesel*,[92] *bwecal*,[93] *bwaichelle* 'fillette',[94] *basselle (bäsel)*,[95] *baisselle*,[96] *bachelette*,[97] *braixelette*,[98] *bacelote*,[99] *bacelatte*,[100] *braichelatte*,[101] *bassotte*,[102] *bèssotte*,[103] *bassatte*,[104] *besotte*,[105] *bosote, bozette, bozonette, besot*,[106]

[76] Gascogne.
[77] Puitspelu, 'Un conte en patois lyonnaise', Rev. de phil. fr. et de lit., 1.30-55.
[78] Pauli, op. cit., Alsace and Vosges dépts.
[79] Ibid., see note 78.
[80] Ibid., Aubure (H.-Rhin) and La Poutroie (Alsace).
[81] Ibid., La Poutroie (H.-Rhin); also: *bâsse, beyèsse*, FEW 1.196.
[82] Ibid., Pexonne (Meurthe-et-Moselle).
[83] Ibid., Belfort (Aude).
[84] Ibid.; Pauli's general statement is that the older meaning of 'servante' remains in Picardie, in the south of Luxembourg, and in the north of the dépt. of Meurthe-et-Moselle.
[85] Ibid., Luxembourg (south).
[86] Ibid., Meurthe-et-Moselle.
[87] Ibid., Gascogne.
[88] Ibid., Liège and Malmédy (in the north of Wallonie). *E* as in note 61.
[89] Ibid., Charleroy and Namur (Belgium).
[90] Ibid., Gedinne (Belgium). *C* bears a short straight stroke crossing its arc near the base, beginning in the center of the enclosure made by the letter and extending in a southwesterly direction.
[91] Ibid., St. Hubert (Belgium), also pronounced *bwēsal* (*e* as in note 42).
[92] Ibid., Bastogne (Belgium).
[93] Ibid., Haybes (Ardennes). *C* as in note 90.
[94] Ibid., Ardennes.
[95] Ibid., environs of Metz to Courcelles, Pange, Falkenberg to the east, and in the dépt. of Meurthe-et-Moselle; also Lunéville (Meurthe-et-M.), Rémilly (Moselle), St. Quirin (Moselle), Lorraine, Meuse, see FEW 1.196-7. In the term between parentheses *e* is as in note 61.
[96] Ibid., occurs in same locations as in note 95.
[97] Ibid., Picardie, French Flanders, and Hainaut. Pauli cites similar sous-diminutives *baisselete* and *bachelete* as being used by La Fontaine.
[98] Ibid., Ardennes.
[99] Ibid., Messin patois.
[100] Ibid., see note 99.
[101] Ibid., see note 99.
[102] Ibid., Lorrain patois.
[103] Ibid., see note 102.
[104] Ibid., see note 102.
[105] Ibid.; Pauli cites this and the three forms following it as being, according to Roquefort, O. Fr. forms. He also remarks that suffixes *-elotte, -elatte, -otte, -atte* are often employed farther to the south.
[106] Ibid., Jung Münsterol (Alsace, near Belfort territory). *E* and *o* each bears a tiny hook directly under it opening toward the top; *s* bears a cedilla symbol directly under it.

bäsnat 560,⁴⁸ *brèhhelate* 560, *brèhhelote* 560, *brexlat* 560,⁴⁹ *brexlot* 560,⁵⁰ *bacelle* 575, *bacelote* 575, *bacelotte* 579,⁵¹ *boèyèsse* 579 'jeune fille',⁵² *béyèsse* 579,⁵³ *béesse* 579,⁵⁴ *boayesse* 579,⁵⁵ *boèyesse* 579,⁵⁶ *bouaïésse* 579,⁵⁷ *bagassia* 579,⁵⁸ *baslot* 589 'fille',⁵⁹ *basot* 589,⁶⁰ *bwayes* 589,⁶¹ *bweyes* 589,⁶² *bwĕyĕs* 591 'fille',⁶³ *bwayes* 591 'fille', *baiçote* 601, *bace* 601 'jeune fille',⁶⁴ *bacelle* 601, *bachelette* 601, *baicelette* 601, *béçate* 601, *béceta* 601, *baichate* 613, 'fille',⁶⁵ *gechotte* 623,⁶⁶ *béssète* 628, *ghèchòte* 628 'jeune fille', *baichotta* 647 'petite fille', *bessaula* 647 'petite fille',⁶⁷ *bechaula* 647, *bressaula* 647, *baichotte* 647, *gaichotte* 647,⁶⁸ *baissette* (*bēsèt*) 667, *baisseta* (*bēstá*) 667,⁶⁹ *baissatte* 667,⁷⁰ *boye* 745,⁷¹ *boetsa* 758 'femme',⁷² *boussa* 758 'femme',⁷³ *bautse* 758 'femme',⁷⁴ *boice* 758 'femme mariée',⁷⁵ *bagasso* 786 'woman of evil life, man

⁴⁸ Villiers-aux-Bois (H.-Marne). Initial *a* as in note 46.

⁴⁹ Messin patois at Vigy. *E* as in note 42.

⁵⁰ L'Isle patois at Maizières and Verny; Pays-Haut patois at Amanvillers and Gorze. *E* as in note 42, *o* as in note 47.

⁵¹ Term of endearment.

⁵² This meaning applies to all terms here listed as from 579.

⁵³ Le Tholy (Vosges).

⁵⁴ See note 53.

⁵⁵ Vagney (Vosges).

⁵⁶ Saint-Amé (Vosges).

⁵⁷ Thiriat (Vosges) and Ventron (Vosges).

⁵⁸ Provençal.

⁵⁹ This meaning applies to all terms here listed as from 589.

⁶⁰ St. Nabord (Vosges).

⁶¹ La Bresse (Vosges). *E* bears an almost closed hook directly under it opening toward the top.

⁶² Le Menil (Vosges).

⁶³ Bloch refers to O. Fr. *baiesse*. Each *e* bears a grave accent directly above the short mark.

⁶⁴ This meaning applies to all terms here listed as from 601. There may be some rapport with the O. Fr. *baceler, bacheler,* 'young man' < Lat. *bacalarii*.

⁶⁵ Used especially to denote the sex of a child that has just been born.

⁶⁶ Evidently a cross between a derivation from *bacelle* and one from *garse*.

⁶⁷ This meaning applies to all terms here listed as from 647. These three terms are in use in Fribourg. Bridel and Favrat cite the Hebrew *bethula* in this connection.

⁶⁸ Possibly influenced by the initial *g* of *garça;* see chapter on *garce*.

⁶⁹ Canton of Neuchâtel (Switzerland). Initial *e* bears a grave accent in both these terms here listed as from 667.

⁷⁰ In Jura Bernois at bishopric of Bâle.

⁷¹ Puitspelu cites the Savoie form *bouille* and suggests that there is some connection with **bagucula*, formed on Celt. *bach* 'little', from which comes **bachgenes* 'young girl'.

⁷² Haute-Savoie.

⁷³ Savoie.

⁷⁴ Suisse Romande.

⁷⁵ Jura Meridionale.

TERMS DERIVED FROM *BACASSA, *BAGASSA, BACHES 53

bouaichelle 478,[15] *braixelette* 478,[16] *boayèsse* 532,[17] *boèïesse* 532,[18] *bèïesse* 532,[19] *boyesse* 532,[20] *bayesse* 532,[21] *beyesse* 532,[22] *bèyèsse* 532,[23] *bèésse* 532,[24] *béese* 532,[25] *bacelle* 532,[26] *bacèle* 532,[27] *basselle* 532,[28] *baicelle* 532,[29] *bêcelle* 532,[30] *bécelle* 532,[31] *bèssatte* 532,[32] *bracelatte* 532,[33] *baicelatte* 532,[34] *bacelotte* 532,[35] *bassatte* 532,[36] *bassotte* 532,[37] *basselle* 540 'jeune fille, servante',[38] *baisselle* 540, *bouaichelle* 540,[39] *baixelette* 540,[40] *bacèle* 560,[41] *bäsel* 560,[42] *bäsēl* 560,[43] *bāsēl* 560,[44] *bacelate* 560 'bachelette, petite fille',[45] *bäslat* 560,[46] *bāslot* 560,[47]

[15] Ardennes.
[16] See note 15.
[17] Vagney (Vosges); the meaning of this and all subsequent terms here listed as from 532 is 'fille'.
[18] Ramonchamp (Vosges).
[19] Le Tholy (Vosges).
[20] Vieuville.
[22] Champdray (Vosges).
[23] Rouges-Eaux (Vosges).
[24] Ban-sur-Meurthe (Vosges).
[25] Mandray (Vosges).
[26] Moyenmoutier (Vosges), Vexaincourt, Cirey, Verdenal (Meurthe-et-M.), Landremont (Moselle), Maconcourt (Vosges), Charmois-devant-Bruyères (Vosges).
[27] See note 26.
[28] See note 26.
[29] Courbessaux, Mailly, Art-sur-Meurthe, Malzéville, Custines, Marainville (all of these in Meurthe-et-M.).
[30] See note 29.
[31] See note 29.
[32] Hergugney (Vosges).
[33] Landremont (Meurthe-et-M.).
[34] Lemainville (Meurthe-et-M.).
[35] Maconcourt, Gelvécourt, Haillainville, Docelles (all these in Vosges).
[36] Gircourt-les-Viéville (Vosges).
[37] Vittel (Vosges).
[38] These meanings apply to all terms here listed as from 540.
[39] Marne.
[40] Ardennes.
[41] Zeliqzon explains that the young girl is at first *bacelîn*, then *bacèlon*, then *bacelate*, finally *bacèle*.
[42] Messin patois at Vigy; l'Isle patois at Maizières and Verny; Pays-Haut patois at Amanvillers and Gorze; Fentsch patois at Fontoy; Nied patois at Frécourt, Sorbey, and Remilly. *A* bears a long mark over the diaeresis; *e* bears an open hook directly under it opening to the right.
[43] Saunois patois at Dieuze, Château-Salin, and Ommeray. *A* bears a long mark over the diaeresis.
[44] Vosgien patois at Gondrexange, Lorquin, and Réchicourt.
[45] Term of endearment.
[46] Messin patois at Vigny; Nied patois at Frécourt, Sorbey, and Remilly. *A* bears a long mark over the diaeresis.
[47] L'Isle patois at Maizières and Verny; Pays-Haut patois at Amanvillers and Gorze. *O* bears an open hook directly under it opening to the right.

Chapter XXVI

TERMS DERIVED FROM *BACASSA, *BAGASSA, BACHES

Derivatives from *bacassa or *bagassa 'maid',[1] or from baches,[2] are common in the Romance tongues and especially in French and Provençal. There are also many diminutives of these in use.[3] Such variants as basse, bayesse, baisselle, basselotte, bachelotte, of which the oldest meaning is 'servante' are common in the patois of eastern France from Wallonie to Jura.[4] In the Neuchâtel region and that of Jura bernois these words are often used to designate 'jeune fille' or 'fillette'.[5] In O. Fr. there existed the terms baiasse, baiesse, baasse, baesse, baisse, basse meaning 'servante, femme de chambre'; in O. Prov. bagassa 'woman of evil life'; in Mod. Fr. bagasse, which is obsolescent, derived from the above O. Prov. form; the Mod. Prov. bagasso has the same meaning as the old form.[6] The following are the occurrences in the dialects and patois with the common meanings of 'fille, fillette, petite ou jeune fille': bouaichelle 2, bassel 38, bwêsal 136, bwesal 137,[7] basal 137,[8] bwesel 137,[9] bwecel 137,[10] baisselette 142, baisselle 142, bachelette 164 'marriageable young girl', baucelle 164 'young person of feminine sex',[11] bacelle 478 'fille',[12] baisselette 478 'fille',[13] bauchelle 478,[14]

[1] REW 861; Romania 23.325, n. 1; Kt. 1131; Gam., p. 65, cites Arab. bâgiza, fem. of bagiz, 'joined together by welding, obscene'. Eguilaz 331 cites baguiyya 'wench, maid'.

[2] Kt. 1140; from Welsh bach 'small' plus Gr.-Lat. suffix -issa.

[3] Pauli, Ivan, 'Enfant', 'Garçon', 'Fille' dans les langues romanes, Lund, 1919, pp. 156–160. He remarks that the diminutives from *bagassa, etc. are numerous and generally signify 'jeune fille', whereas the simple word signifies 'fillette'. Gam., p. 65, cites baisselette as a 13th Cent. O. Fr. form from baisselle as a diminutive of 12th Cent. baiesse and baisse 'maid, maid-servant', Norm. basse, Guernesey baisse, H.-Jura batze, boitze 'girl, lass'.

[4] Ibid., 156.

[5] Ibid., 156.

[6] Ibid., 156–7; also O. Fr. and Mid. Fr. baisselete, Mid. Fr. and Mod. Fr. bachelette, see FEW 1.196–7. Arthur Langfors, Romania 62.394–5, gives baiesse as a form or variant of abeesse 'abbess'.

[7] Arville (Meuse).

[8] Champlon (Meuse).

[9] Porcheresse (Charente) and Anloy.

[10] Haut-Fays and Opont.

[11] The same as Fr. baicelle 'servante'.

[12] Ardennes and Namur.

[13] See note 12.

[14] Rilly-aux-Oies (Ardennes).

TERMS DERIVED FROM FILIA

In Mid. Fr. occurs *fillaude* 'little girl',[23] *feillaude*,[24] *fillaude*,[25] *felaude* 353.[26]

The following signify 'prostitute': *fillasse*,[27] *filase*,[28] *fihasso*,[29] *filhasso* 789.

There is a miscellany of such formations many having a pejorative significance: *fiotasso* 969 'fille évaporée', *fillasse* 373 'grande fille'; *fillon* 777 'petite fille', *fillas(so)* 940 'fille puissante ou de haute taille', *fiyaz* 204 'fillettes ou jeunes filles évaporées prises collectivement'; *fillasse* 'petite fille à allures de garçon, grande et grosse fille sans grace',[30] *filasuno* 'fillet rondelette, petite dondon',[31] *fillondras* 'grande fille mal mise, malpropre',[32] *fillastre* 'grande fille mal mise',[33] *filū* 'les filles en général',[34] *fiun* 'la généralité des filles d'un endroit (en sens narquois)',[35] *fihan* 'les filles (terme de dénigrement)'[36] *fihan* 838 'les filles en général', *filhan* 789 'jeunes filles en général', *filhandran* 789 'les filles en général', *filhas* 789 'large girl, virago', *filhatas* 789 'fille géante', *figlie* 779,[37] *figlietta* 779,[38] *fillea* 334,[39] *fẏilya* 978 'fille'.[40]

[23] Used by Brantôme. FEW loc. cit.
[24] Poitevin, FEW loc. cit.
[25] Saintonge, Aunis, and Centre (ALF 523, 525, 533, 535, 536), FEW loc. cit.
[26] There is an echo vowel y between l and a.
[27] Mid. Fr. and Mod. Fr., Lyonnais.
[28] Mid. Dauph.; in source final e is represented as phonetic symbol French mute e; l as in note 16.
[29] Mod. Prov.
[30] Lyon.
[31] Dauph.; in source u is printed as U.
[32] Aveyr., Masc.
[33] Aveyr., here a specialized meaning; this formation occurs in many parts of the country meaning 'daughter-in-law', as *felatre* 353 (with echo vowel y between l and a) and *fillatre* 364.
[34] B.-Dauph.
[35] Languedoc.
[36] Prov.
[37] Equivalent to Fr. *fille*, this has evidently been taken directly from It. *figlio, figlia*; see Pauli, op. cit., 99–102.
[38] Equivalent to Fr. *fillette*, evidently from It. *figlietto, figlietta*; see Pauli, loc. cit.
[39] Poetical.
[40] With sense of kinship; i bears short straight stroke directly under it slightly inclined toward lower right; l bears a small arc directly under it opening toward top; a bears a small circle directly above it.

Chapter XXV

TERMS DERIVED FROM *FILIA*

A diminutive formed by reduplication and used as a term of endearment is used in Mod. Fr. and in the dialects: *fifille*,[1] *fifil* 204, *fifille*,[2] *fèfèye*,[3] *fefey*,[4] *fefoey*,[5] *fèfeuye (fefœy)* 560,[6] *fèyèye (feféy)* 560.[7]

Variations of *filuno* occur with the general meaning of 'girl': *filuno*,[8] *fillou*,[9] *filhouno*,[10] *filuno*,[11] and on the ALF map no. 1477 at numbers 705, 813, 817, 844. There is a form *fedetta* 647,[12] for which see also number 969 on map no. 1477 of the ALF.

The following have the meaning of 'daughter-in-law': *filiát* 756, *fiyāda* 956,[13] *xilyado* 980,[14] *fihado*,[15] *filado*,[16] *fiyáda*,[17] *fillada*,[18] *fiiada*,[19] *filado*,[20] *fyilyáda*,[21] *filhado*,[22] *filiado* 865.

[1] Mod. Fr. 'dear little thing', see FEW Lief. 22.516, belongs to familiar speech, found in the Latin inscriptions of Gaul as *fifilla*.

[2] Anjou (now dépt. of Maine-et-Loire and southern parts of dépts. of Mayenne and Sarthe), FEW loc. cit.

[3] Louvain, FEW loc. cit.

[4] Namur (Belg.), FEW loc. cit.

[5] Metz (Moselle), FEW loc. cit.

[6] Sounois pat. at Dieuze, Château-Salin and Ommeray; here it signifies 'preferred daughter'.

[7] Nied (Moselle), 'preferred daughter'.

[8] Cordéac (Isère), FEW loc. cit.

[9] Aveyr., FEW loc. cit.

[10] Coùx (Ardèche), FEW loc. cit.

[11] M. Dauph., H.-Loire, Puy-de-Dome; in all these cases it is equivalent in meaning to Fr. *fillette*, FEW loc. cit.

[12] '(Jeune) servante'; Pauli, Op. cit., p. 323, says that *fedo* (of which *fedetta* is a dim.), meaning 'garce' in Prov., is doubtless the same as Prov. *fedo* 'brebis'. *Fede* 667 'girl' also occurs (bears a dot over each *e* and one under *d*).

[13] Bears an acute accent above the long mark over *a*.

[14] In the source *x* is printed as an italicized capital.

[15] Mod. Prov., FEW loc. cit.

[16] Ardèche, FEW loc. cit.; *l* bears a short straight stroke across center of bar, crosses at right angles; the tip of stroke is turned down on left and up on right side.

[17] H.-Loire, FEW loc. cit.

[18] Velay (H.-Loire), FEW loc. cit.

[19] See note 18; second *i* bears small arc directly under it opening toward bottom.

[20] See note 17; *i* bears short arc under it opening toward the right; *l* as in note 16.

[21] Final *a* bears small circle directly above it. Vinzelles (Puy-de-Dôme), FEW loc. cit.

[22] Perigord dial. (in Dordogne and northerly Lot-et-Garonne), FEW loc. cit.; see also ALF map no. 1477 *bru*, nos. 805, 811, 812, 804, 813, 814, 815, 806, 807, 808, 809, 821, 824, 825, 827, 829, 833, 842, 912, 913, 914, 915, 917, 919, 921, 922, 924.

'women in general',[44] *femrey* 'women in general',[45] *fennun* 'women in general',[46] *feum'rèye* 'women in general',[47] *fomerèye* 'women in general',[48] *famail* 'band of women',[49] *fennée* 'woman',[50] *fénôle* 'woman',[51] *fennon* 'my little woman',[52] *fenõ* 'little thick-set woman',[53] *femmele*,[54] *femmelette* 'woman weak in body and spirit',[55] *feum'lète* 'woman weak in body and spirit',[56] *fœmlot* 'woman weak in body and spirit'.[57]

[44] Ibid., Puisserguier (Hérault, Béziers).
[45] Ibid., Malmédy. Each *e* bears a hook directly under it opening toward the right.
[46] Ibid., see note 44.
[47] Ibid., Louvain.
[48] Ibid., Verviers (Belgium, Louvain).
[49] Ibid., 16th Cent. term, 'troupeau de femmes'.
[50] Ibid., Poitiers.
[51] Ibid., Geneva.
[52] Ibid., Albertville (Savoie); 'ma petite femme, terme d'amitié'.
[53] Ibid., Montrevel (Isère, La Tour-du-Pin, Virieu). *E* is printed as the international phonetic symbol for French mute *e*.
[54] Ibid., Mid. Fr., 16th Cent. term.
[55] Ibid., Fr., in use since 14th Cent.
[56] Ibid., Louvain.
[57] Ibid., Pays-Haut. This symbol and the following *o* each bears a hook directly under it opening toward the right.

'fat and ugly woman', *hemnasse* 788,[19] *hemnete* 788,[20] *hemnine* 788,[21] *hemnou* 788,[22] *fènot (féno)* 627 'little woman', *femasso* 789 'large woman, fat woman, ugly woman, mean woman',[23] *fremasso* 789,[24] *fenoulet* 810 'very small woman', *fenassa* 810 'fat woman', *fenouna* 810 'small woman', *fenach* 810 'fat woman', *fremouno* 850 little woman', *fremeto* 850 'woman of little importance', *fenneto* 865 'small woman', *fennun* 940,[25] *fennasso* 940 'large woman', *fenièu* 962 'widow', *fennoto* 968 'woman', *finnona* 972 'small woman', *finnassa* 972 'large woman',[26] *fennoto* 962 'woman', *fennāsso* 991 'large or fat woman', *fennoto* 991 'little woman', *femio* 1022 'female', *hemnot* 1035 'femme sans relief',[27] *fenneto* 1044 'woman', *hemnete* 1094 'little woman', *femmete* 'woman of little importance',[28] *femneta*,[29] *fomat*,[30] *fomot*,[31] *fŏnnate (fõnat)*,[32] *fennetta* 'little woman',[33] *fenneto* 'little woman',[34] *fremeto* 'little woman,[35] *fenneto* 'little woman',[36] *fremasso* 'large woman',[37] *fẽnāso* 'large woman, prostitute',[38] *fẽnālo*,[39] *feme* 'femelle, femme',[40] *feme* 'femelle',[41] *femouno* 'little woman',[42] *fenun* 'women in general',[43] *fénnat*

[19] See note 12.
[20] Béarnais diminutive of *femno*.
[21] Same as note 17.
[22] Same as note 17.
[23] See E. L. Adams, Word-Formation in Provençal, New York, 1913, p. 140: 'Suffixes -*as*, -*asa* are generally from Latin suffixes -*aceus*, -*acea;* this suffix gives an idea of quantity to the simple word, or greatness in size, and the depreciative force'.
[24] This term and the ones here listed from 789 are synonyms of *femasso: fennasso, hennasso*.
[25] See note 12.
[26] The first *a* is in italics type.
[27] Masculine gender.
[28] See FEW lief. 21, p. 449f; O. Fr., 13th Cent. and 14th Cent. This and the undefined terms following have identical meanings.
[29] Ibid., Prov.
[30] Ibid., Metz, Nied, Saunois. *O* bears a hook directly under it opening to the right.
[31] Ibid., Isle. Each *o* bears a hook directly under it opening to the right.
[32] In the printed source from which taken there is a dot between the *n*'s placed equidistant from top and base of the letters.
[33] Ibid., Swiss.
[34] Ibid., Mod. Prov.
[35] Ibid., Prov.
[36] Ibid., Toulouse.
[37] Ibid., Prov.
[38] *A* bears an acute accent above the long mark indicated in the text.
[39] Ibid., Middle Dauphinois; means 'women in general'. *A* bears an acute accent above the long mark indicated in the text; *l* bears a short straight diagonal stroke through it from upper right to lower left.
[40] Ibid., O. Pr.; adj. and masc. subst.
[41] Ibid., Rouergue (Aveyron).
[42] Ibid., Mod. Prov.
[43] Ibid., H.-Dauph., Grenoble, B.-Dauph.

Chapter XXIV

TERMS ON THE SAME STEM AS *FEMME*

There are a number of diminutives and augmentatives on the same stem as *femme*. Most of these have been formed by adding some dialect or patois variation of the usual French suffixes: *-ail, -as, -et, -ole, -ette, -onne,* etc.[1] In the dialects and patois occur: *famelotte* 229 'woman weak in body and spirit',[2] *femreie* 57,[3] *femmotte* 373 'little woman', *feumrèie* 105 'married woman, unmarried woman, widow', *fumèrie* 105 'married woman' etc., *fonnelòte* 448 'woman weak in body and spirit',[4] *fomat* 560 'femmelette',[5] *fomot* 560 femmelette',[6] *fannelote* 426 'little woman', *fennetta* 647 'little woman', *fennon* 647 'my little woman',[7] *fonnotte* 620 'woman of little importance', *fénole* 667 'woman',[8] *féniole* 667 'woman',[9] *fenol* 667 'woman',[10] *feñol* 667 'woman',[11] *fénôle* 703 'young woman',[12] *fenotte* 745 'woman',[13] *fennotta* 759 'little woman', *fenon* 763 'young bride', *fenō* 772 'little woman',[14] *fenum* 775,[15] *fennota* 777 'little woman', *femnaron* 788 'little woman, nice little woman',[16] *femneto* 788 'little woman, nice little woman',[17] *femnaroú* 788,[18] *femnasso* 788 'fat and ugly woman', *fremas* 788

---

[1] See F. Brunot, La pensée et la langue, Paris, 1922, p. 657ff.
[2] Since 14th Cent.
[3] Collective term of scorn for curious women.
[4] See note 2.
[5] Messin patois at Vigy; Nied patois at Frécourt, Sorbey, Remilly; Sounois patois at Dieuze, Château-Salin, Ommeray. *O* bears a hook directly under it opening toward the right.
[6] See note 5. Each *o* bears a hook directly under it opening toward the right.
[7] Term of endearment.
[8] Suisse Romande, used in a general sense, not pejorative.
[9] See note 8.
[10] *E* and *o* each has a hook directly under it opening toward the right.
[11] Same as in note 10.
[12] Canton of Geneva.
[13] See note 7.
[14] *E* should be the international phonetic symbol for Fr. mute *e*.
[15] Also syn. *femnum, femná, femnage, femelan, femelun,* as well as *fenum,* signify 'woman in general.' All are derived from *femina* except the last two, which are from *femella,* as is the augmentative *fremalas.*
[16] 'Gentille petite femme'.
[17] See note 13.
[18] This has the same meaning as the two preceding terms. From 788 other synonyms of these terms are: *femnoto, femnou, femnouno, femnouneto, femnoutil, feneto, fremeto, fremouno.*

CHAPTER XXIII

SALOPE

A number of etyma have been proposed for *salope*. Körting suggests Dutch *slap* 'slack, loose'.[1] Schuchardt gives the same etymon and says that the word is equivalent to the German *Schlump* 'slut', and *Schlampe* 'slattern'.[2] Tobler denies that there is any connection with the French *sale*, saying that it is highly improbable that *sale* gave *salot* and *salotte* and these became *salop* and *salope* by an analogy with *galoper*. He says that the form *salope* did not appear until the 17th Cent., and he connects it with *marie-salope* 'name of a type of boat made for carrying slime and filth from a port'. Though he suggests no etymon himself, he remarks that the word is doubtless of Germanic origin, probably Dutch.[3] Bloch derives 13th Cent. French *sale* from Old High German *salo* 'trouble', M. H. G. *sal*, mentioning *salope* as a derivative.[4] Diez proposes the English *sloppy*.[5] As to its meaning, Cotgrave cites it as 'a sloven, or slut', from Orléannais.[6] The word is today considered a popular expression, and means 'slut' or 'trollop' just as it formerly did. Littré qualifies it slightly: 'one who is dirty or improper'; fig. and as an insult, 'a woman of evil life'. The word is used both as an adjective and as a noun. Among its occurrences are: *salop* 38, *salop* 105, *salope* 579,[7] *salope* 151, *salop* 789, *salope* 789, *salopo* 789, *solopo* 813, *salopo* 828, *salópa* 978,[8] *salopo* 987, *salopo* 1023 'woman of evil life',[9] all of feminine gender. In addition there are the derivatives *salopiaud* and *salopiaude*, *saligot* and *saligote*,[10] *salopiot* and *salopiote*, all from 667, which are used in the popular language as equivalents for the popular French *salaud* and *salaude*,[11] 'sloven, slut', and *saligaud* and *saligaude* 'sloven, slut, dirty thing', used as nouns.

[1] Kt. 8804.
[2] Schuchardt, Salope in *Z* 21.130 (1897).
[3] Tobler, Salope, in *R* 25.623-4 (1896).
[4] Bloch, 2.251.
[5] Dz. 675.
[6] Cotgrave, see *salope*.
[7] *A* alone is in italics type in source.
[8] Each *a* is in italics type and bears a tiny circle above it; *o* bears a short straight, almost vertical stroke under it directed slightly toward lower right.
[9] Here Durrieux cites the Gr. ψόλος, σαλάβη, and also lists the Fr. meanings 'saleté, érasse, la Venus prostituée'.
[10] *A* alone is in italics type in source.
[11] Tobler, loc. cit., says these forms are of the 16th Cent.

590,[12] *putèn* 590,[13] *puta* 632, *poutan* 647, *puta* 647, *putan* 647, *putăhăna* 675, *puta* 686, *puta* 715,[14] *puta* 759, *putarâssi* 759, *putéin* 759, *putaio* 789,[15] *putatho* 789,[16] *putan* 789,[17] *putano* 789,[18] *putanello* 789,[19] *puteto* 789,[20] *putarrou* 789,[21] *putasso* 789,[22] *puto* 789,[23] *puto* 921, *puto* 970, *putanèiro* 980, *putana* 1023, *putane* 1023, *putharasso* 1023 'incorrigible prostitute', *puthâsso* 1023, *putho*.[24]

[12] *E* bears the international phonetic symbol of nasality above the grave accent.
[13] Same as note 12.
[14] Suisse Romande.
[15] Languedocien and Gascon, meaning 'putains en général'.
[16] See note 15.
[17] Dialect of the Alps.
[18] See note 17.
[19] Béarnais,—the meaning here is, of course, diminutive.
[20] See note 19.
[21] Languedocien,—diminutive.
[22] Augmentative.
[23] Mistral gives this as the general Provençal term and says that in addition to its usual meaning it also signifies 'round flat cheese of a poor kind' in Forez.
[24] Durrieux refers us to Gr. πόθος.

CHAPTER XXII

PUTE, PUTAIN

Putain is the objective case of the Old French *pute*, which is still listed in dictionaries as the feminine of the Old French adjective *put*, 'stinkard, wicked', from which came the meanings 'evil', 'dirty', etc., having in the east of France today the significance of 'laid'.[1] The etymon of these two forms is generally accepted as *pŭtĭdŭs,-a,-um*, a Latin adjective meaning 'stinking',[2] derived itself from the Latin verb *pūteō,-ēs,-ēre*, 'to be rotten, spoiled, corrupted',[3] though one authority would trace it to *pŭtus,-ī*, 'small boy, child'.[4] These French forms are of the 12th Cent.[5] The *-ain* ending has been explained through analogy with the French *nonne* and *nonnain*.[6] In Old French the meaning of the noun was 'prostitute', and there are a number of variations existing in Old French of the adjective, *put*.[7] Today the words exist individually, both *pute* and *putain* meaning 'prostitute',[8] gross and impolite terms having the extended meaning of 'fille, femme débauchée'.[9] Cotgrave gives a number of colorful meanings for the word: 'whore, punke, drab, flurt, strumpet, harlot, cockatrice, naughty pack'.[10] Among the dialect occurrences of the two forms are: *putin* 105, *pitaînn* 105, *putin* 151, *putain* 522,[11] *putē* 541, *putain* 579, *putè*

---

[1] Bloch 2.195.

[2] Dz. 662, Gam. 726, Kt. 7578. The last postulates a form **pūtĭdānă, -am*, based on *putıdus*.

[3] Ernout et Meillet, Dict. Etym. de la langue latine, Paris, 1932, p. 788–789.

[4] Ibid., p. 790. Foerster in *Z*, 3, 1879, p. 565–566, contributes nothing of importance to the study of *pute*'s derivation except that he draws an analogy from *nĭtidus, neto, nete, net, putidus, puto, pute*. G. Paris in *R*, 9, 1880, p. 333–334, adds little other than that *puta* would have given *peue* by natural phonological development and not *poue* as Foerster suggested, explaining that free accented *o* gives *eu*, and that close accented *o* gives *u*, written *ou*.

[5] Gam. 726.

[6] Bloch 2.195; Littré, op. cit. 3.1391 and 3.746. The latter remarks that Jules Quicherat has found proofs of direct forms in *a* and obliques in *ana* in Merovingian proper names brought about through the influence of the weak Germanic declension.

[7] Godefroy, op. cit. 6.472 gives the adj. variations *put, pust, puit, puet, peut, pout, pot*, having the meanings of 'sale, infect, mauvais, méchant'.

[8] Passy and Hempl, International French-English and English-French Dict., New York, 1904, p. 483.

[9] Littré 3.1391.

[10] Cotgrave, op. cit.

[11] Goffart says this is used like *garce* in oaths and apostrophes without bearing the insulting and injurious meaning here that it ordinarily has elsewhere.

CHAPTER XXI

PUCELLE

A number of etyma have been suggested for this word. *$P\breve{u}ll\breve{\imath}c\breve{e}lla$, dim. of *pulla* (fem. of *pullus*, 'young animal') has been suggested by some,[1] and objected to by others, due to the difficulties encountered in the development of $\breve{u}$.[2] *$P\bar{u}licella$, dim. of $p\bar{u}lice$, 'flea', becoming general after first being used as a term of endearment, is phonologically possible but offers obvious semantic difficulties.[3] *$Puellicella$, dim. of *puella*, 'girl', is semantically the most sound of the lot, but the formation is impossible.[4] *$Pulcher-lla$, adj. 'beautiful' plus dim. suffix *-ella*, is not likely.[5] *$P\bar{u}ricella$, 'maid-servant', dim. built on *puer* (*puršela* in Oberwaldisch), with *r* lost by assimilation would come into Old French as *pulcelle*.[6] The meaning of the word was 'young girl';[7] 'virgin', 'unmarried girl', in Old French (the last meaning by extension).[8] *Pulicella* has been attested in *Lois Barbares* of the 6th Cent.[9] In Old French the following terms are attested: *pulcella, pulcela, pulcele, puicelles, pucele, pucheles, poucelle*.[10] Cotgrave defines the term as 'virgine, maid, girle, damsell, mother', and contributes the form *puceale* (masc. *puceal*), 'virgine, maiden, or maidenly, virgin or maiden-like', used as a subst. or adj.[11] Among the present-day dialect and patois occurrences are: *pucelle* 29, *puchéle* 203, *pussalla* 647, *pusala* 675, *pieucelo* 786, *piéucello* 789,[12] *piucelo* 917, *piucèlo* 1045, *puncele* 1094.

[1] REW 6819, Dz. 258, Pauli 86.
[2] Foerster in *ZRP* 16.254–255.
[3] Kt. 7517; Foerster in *ZRP* 16.254–255.
[4] Gam. 724, Kt. 7506.
[5] An unpublished etymology by C. M. Hutchings.
[6] Jordan in *ZRP*, 43.708.
[7] Littré, 3.1379.
[8] Godefroy, op. cit. 10.444.
[9] Bloch 2.192.
[10] See note 9.
[11] Cotgrave, op. cit., see *pucelle*.
[12] Mistral 2.572, includes a number of forms used as masc. adjs. and masc. substs.: *piéucèu, piéusèu* (Marsellais); *piéusèc* (Dauphinois); *piéucèl, piéusèl* (Languedocien); *pieucel, piucel, piussel, piusel, piuselh, piuzel, pieuzel, puissel, pucel* (Old Provençal). One would suppose that each of these had corresponding fem. forms. The fem. form he lists from lit.:

'Si un ome jouve, piéucèl,
Espouso uno filho piéucello
Li vendra mal a la maissello'—P. Ducèdre. Mistral gives an etymon for these masc. forms: *pusillus*, meaning 'very little, very small, petty insignificant' as an adj., and as a subst., 'a very little, trifle', from *pūsus*, 'boy'. The fem. form of this adj. could have been the etymon for *pucelle*.

43

Chapter XX

MOLLER

In addition to *moller* 767, *muller* 767 and *moiller* 767 are derived from CL *mŭliĕr*, *mŭliére* (or *mulierem*) meaning 'woman' or 'female',[1] brought into the vulgar language of Dauphiné in about the 13th Cent. Had the word developed into French the normal sequence would have been: *mŭliĕrem* > **mŭliĕre(m)* > **moliere* > *moiller (mouiller)*, *ŭ* becoming *o* in VL, and *li* becoming palatal *l*, (spelled -*ill*- in Low. Rom.).[2] The standard Provençal forms are *molher* and *molhér*; the Catalan, *muller* and *móller*. Other occurrences in French, Provençal and bordering dialects and patois are: *moillier* 478 'femme épouse' Namur, *moglie* 779 (identical with the regular development in Italian, *molher* 786, *moulhe* 786, *mouie* 786, *mouillet* 828[3], *moller* 837, *mouié* 839, *moulhé* 891, *mouillé* 915, *mouellé* 917, *mouliè* 921, *mouilhè* 921, *mouihé* 1022, *moulhè* 1022, *mulhè* 1022, *molher* 1022, *mouilhe* 1023, *moulhè* 1045, *mulhê* 1057, *mŭlē* 1068,[4] *mŭgēr* 1068,[5] *mouillé* 1047, *moulhè* 1088, *moulhé* 1093, *moulhe* 1094, *molher* 1094, *mouille* 1100, *molher* Hem., *molheir* Hem., *moilher* Hem., *molier* Hem.,[6] *moillier* 478, *miglia* 758. In every case the meaning is that of 'married woman' with other possible meanings of 'wife', 'woman', 'female'. To be noted above are the regular development of *li*[7] to *gl*, as in Old Italian, in 779; *li*[8] to *g*, as in Old Spanish, in 1068; the regular development in Provençal of *li*[9] to *lh*, as in 891, etc.; and such arrested development as is shown by *molier* 478 and 917.[10]

[1] REW 5730, Kt. 6352.
[2] The *i* of the *li* combination in both starred forms and where the *li* is cited separately bears under it an arc-shaped diacritical mark opening toward the bottom. The first *e* in each starred form bears an acute accent as well as the short mark indicated in the text. The *o* in the second starred form bears a dot under it.
[3] 'Woman, wife'.
[4] The *l* bears an arc-shaped diacritical mark under it opening toward the top; the *e* bears an acute accent above the long mark indicated in the text and also bears under it a straight stroke inclined slightly toward lower right.
[5] The *e* bears an acute accent above the long mark indicated in the text and also bears under it a straight stroke inclined markedly toward lower right.
[6] The reference is to F. Hemmann, Consonantismus des Gascognischen bis zum Ende des dreizehnten Jahrhunderts; Cöthen, 1888.
[7] The li is marked as explained in 2.
[8] See note 7.
[9] See note 7.
[10] Cf. also Guarneiro in Rendiconti del R. Istituto Lombardo 48.704, Milano; and Antoine Thomas in Romania 42.414; also Archivio glottologico italiano 21.29.

CHAPTER XIX

JASASSE, JAZRESS

Jasasse 214, *jazress* 105 are evidently derived from the Fr. verb *jaser* 'to prate, chatter', which was at first applied to the babblings of birds such as the jay, magpie, parrot, blackbird, etc., and then came to be applied to children who were just learning how to talk and to men who speak at random.[1] *Gazouiller*, 14 Cent. form, a diminutive from *jaser*, has undergone the same evolution, passing successively from the songs of small birds to the babbling of children and the murmurings of streams.[2] O. Fr. used *gaziller*, 13th Cent. form, and Prov. has a form *gazalhar*, both being onomatopoeic diminutives of an early *gazar* (Auvergne *gasá*), O. Fr. *gazer*,[3] or *jaser*, 12th Cent. form, the first applied to 'jay', the last more general. As for the actual etymology of these French verb forms, Diez takes them from O. Norse *gassi* 'gander', Littré from Bret. Celtic *geiz* 'warbling (of birds), babbling (of brook)', onomatopoeic parallel to the French and independent of the French, both going back to the same imitative source.[4] Meyer-Lübke derives the words from *gas* 'to chatter, prattle' (onomatopoeic, O. Fr. *jas*, *gas* 'idle talk, gossip, deception, fraud)'.[5] *Jasasse* 214 and *jazress* 105 both mean 'talkative woman'.[6]

[1] Lazare Sainéan, Les sources indigènes de l'étymologie française, 2.46 (Paris, 1925).
[2] See note 1; also Bloch 1.330: *gazouiller* attested in Oresme.
[3] See note 1.
[4] Cited by Sainéan, see note 1.
[5] REW 3696.
[6] See also Stimming in ZRP 30.595.

Chapter XVIII

JACASSE

This is a derivative from the Fr. verb *jacasser* 'to babble, cry' (of the magpie), a 19th Cent. form, and represents a crossing of Norman *agasser*, which gave *agace*, with the dialect form *jaque* 'magpie' in Bresse-Châl. (Sav. *jaquette*), the same form meaning 'jay' in Centre, Burgundy, Lorraine, Anjou, Champagne, Dauphinet regions.[1] For Jura, Vienne, and Marne, *jaquot* is attested, being very much the same as *Jacques*, a dialect development of names for the 'jay', such as *Richard, Colas, Charlot, Gautereau*.[2] To derive *jacasser* from **gaccus* (**gâcus*, a side form), the supposed root for *geai* is unlikely,[3] as Gam. remarks.[4] *Agasser* is derived from O. H. G. *agāza*, a term of endearment,[5] from O. H. G. *ago* 'magpie'.[6] Some dialect and patois occurrences are: *jacasse* 330 'woman of a talkative character',[7] *jacasse* 353 'talkative', *jhacassa* 373 'talkative woman', *jacasse* 395 'said of a girl who talks incessantly of things having no interest for her listeners', *jacasse* 632 'one who has a contrary character'. *Javasse* 356 'talkative', *racasse* 384 'talkative woman', *jagouasse* 398[8] 'primarily, celandine (a plant), secondarily, talkative woman', and *jabiâssi* 759 'thrush, fig. tall woman' are probably connected with the above, though the development of such consonantal replacements is not clear.

[1] Gam. 534; Bloch 1.395 dates *jacasser* 1835 and gives *coasser* as a parallel development.
[2] Gam. 534.
[3] Nigra, Romania 31.518.
[4] See note 2.
[5] REW 275.
[6] Bruinier, Zeitschrift für vergleichende Sprachforschung auf dem Gebiet der indogermanischen Sprachen 34.350.
[7] Ménière refers to O. Pr. *agasse* 'magpie'.
[8] Jaubert refers to an attested form *jagouasser*.

CHAPTER XVII

HOR

This word meant 'prostitute, woman of evil life' and was a Norman form derived from Old Norse *hōra* 'whore'.[1] The standard Old French *holiere*, 'dissolute woman', was derived from the same source, as were also the forms *houilliere, houliere, houriere*, as well as a large number of masculine forms.[2] Old High German *hôra* and *hourâ* have been given as etyma, though they are themselves probably derived from the Old Norse.[3] Occurrences are: *houre* 2, *hore* 57,[4] *houre* 143, *hore* 217,[5] *hoūre* (*hūr*) 560. In Modern French the term does not exist.

[1] REW 4177.

[2] Godefroy IV, 486 487: *holier, holer, houlier, horier, hourier, huler, hourleir, houllier, houiller, hoilier, hærrier, hurier, hoilstier, ellier, erlier*, 'man who frequents prostitutes, dissolute courtier, pimp.' Fem. forms attested: 'Ce font hourieres et hourier' (Jacq. d'Amiens, Remede d'Amor, ms Dresden, f°20ʰ); 'De faire defendre . . . jus de dés, ne hostellent femmes ne houllieres, le jour et le nuyt Sainte Berthe (1507, Prév. de Doulens, Cout. loc. du baill. d'Amiens, II, 77, Bouthor.)

[3] Diez 616; Shorter Oxford Dict., Oxford, 1933. Vol. II, 2423.

[4] Body remarks that this is out of use.

[5] Le Héricher suggests a connection with O. Fr. *gore*, 'sow'. 'Maintes femmes de bourdel ne font leur pechié fors que par povreté, ou pour ce qu'elles furent deceues par mauvais conseil de houlieres et de mauvaises femmes' (Liv. de Chev. de la Tour, p. 255, Bibl. elz.).

CHAPTER XVII

HOR

This word meant 'prostitute, woman of evil life' and was a Norman form derived from Old Norse *hōra* 'whore'.[1] The standard Old French *holiere*, 'dissolute woman', was derived from the same source, as were also the forms *houilliere, houliere, houriere*, as well as a large number of masculine forms.[2] Old High German *hôra* and *hourâ* have been given as etyma, though they are themselves probably derived from the Old Norse.[3] Occurrences are: *houre* 2, *hore* 57,[4] *houre* 143, *hore* 217,[5] *hoūre (hūr)* 560. In Modern French the term does not exist.

[1] REW 4177.

[2] Godefroy IV, 486 487: *holier, holer, houlier, horier, hourier, huler, hourleir, houllier, houiller, hoilier, hœrrier, hurier, hoilstier, ellier, erlier*, 'man who frequents prostitutes, dissolute courtier, pimp.' Fem. forms attested: 'Ce font hourieres et hourier' (Jacq. d'Amiens, Remede d'Amor, ms Dresden, f°20ʰ); 'De faire defendre ... jus de dés, ne hostellent femmes ne houllieres, le jour et le nuyt Sainte Berthe (1507, Prév. de Doulens, Cout. loc. du baill. d'Amiens, II, 77, Bouthor.)

[3] Diez 616; Shorter Oxford Dict., Oxford, 1933. Vol. II, 2423.

[4] Body remarks that this is out of use.

[5] Le Héricher suggests a connection with O. Fr. *gore*, 'sow'. 'Maintes femmes de bourdel ne font leur pechié fors que par povreté, ou pour ce qu'elles furent deceues par mauvais conseil de houlieres et de mauvaises femmes' (Liv. de Chev. de la Tour, p. 255, Bibl. elz.).

Chapter XVIII

JACASSE

This is a derivative from the Fr. verb *jacasser* 'to babble, cry' (of the magpie), a 19th Cent. form, and represents a crossing of Norman *agasser*, which gave *agace*, with the dialect form *jaque* 'magpie' in Bresse-Châl. (Sav. *jaquette*), the same form meaning 'jay' in Centre, Burgundy, Lorraine, Anjou, Champagne, Dauphinet regions.[1] For Jura, Vienne, and Marne, *jaquot* is attested, being very much the same as *Jacques*, a dialect development of names for the 'jay', such as *Richard, Colas, Charlot, Gautereau*.[2] To derive *jacasser* from **gaccus* (**gâcus*, a side form), the supposed root for *geai* is unlikely,[3] as Gam. remarks.[4] *Agasser* is derived from O. H. G. *agāza*, a term of endearment,[5] from O. H. G. *ago* 'magpie'.[6] Some dialect and patois occurrences are: *jacasse* 330 'woman of a talkative character',[7] *jacasse* 353 'talkative', *jhacassa* 373 'talkative woman', *jacasse* 395 'said of a girl who talks incessantly of things having no interest for her listeners', *jacasse* 632 'one who has a contrary character'. *Javasse* 356 'talkative', *racasse* 384 'talkative woman', *jagouasse* 398[8] 'primarily, celandine (a plant), secondarily, talkative woman', and *jabiâssi* 759 'thrush, fig. tall woman' are probably connected with the above, though the development of such consonantal replacements is not clear.

[1] Gam. 534; Bloch 1.395 dates *jacasser* 1835 and gives *coasser* as a parallel development.
[2] Gam. 534.
[3] Nigra, Romania 31.518.
[4] See note 2.
[5] REW 275.
[6] Bruinier, Zeitschrift für vergleichende Sprachforschung auf dem Gebiet der indogermanischen Sprachen 34.350.
[7] Ménière refers to O. Pr. *agasse* 'magpie'.
[8] Jaubert refers to an attested form *jagouasser*.

The tour is supplemented by clear plans and arrows at appropriate points in the library, and, if possible, staff are available to answer questions which may arise. Experience at the University of Delaware has indicated that such tours take about twice as long as a conventional tour[40], but they can be carried out at the student's own convenience, and can also be geared to any particular academic discipline. One British library which has introduced such tours is the Brotherton Library at the University of Leeds.

The use of audio-visual materials, either in addition to or instead of tours is more common, and 10 of the libraries surveyed have an introductory film and 24 an introductory tape/slide presentation. The advantage with these, of course, is that large numbers can be reached at once, although unless a librarian is present no questions can be asked. Audio-visual methods can overcome many of the problems inherent in the conducted tour, in that information can be carefully presented and, for example, close-ups of catalogue entries can be shown. Experience suggests, however, that they are not effective in imparting detailed information. Indeed,

> "there is plenty of evidence that these 'armchair tours' are almost completely ineffective."[41]

For the new undergraduate, their main value seems to be in giving a general impression of the kind of place a university library is and perhaps in bringing to his attention some of the services which are provided.

Lancaster's approach has been to run tape/slide presentations so that the students will see them just before they try to find their own way round the library:

> "An elaborate display was mounted to introduce the new students to the Library. This consisted of a continually running tape/slide commentary on a screen in the entrance hall, with headphones for six readers at a time to listen... It was surrounded by a 'rogues' gallery' of large-sized photographs of subject specialists and reader-services personnel... We hope that... its location in the Library itself will have enhanced the relevance of the information."[42]

Having been shown to a mass audience at the beginning of the year, a tape/slide sequence can be made available for individual viewing in carrels of the type developed by Surrey University. Like self-guided tours, the presentations can then be used at the convenience of the readers. However, all the librarians I spoke to who had such carrels available said that they were very rarely used. This illustrates the crucial problem in reader instruction, the difficulty of actually persuading the reader to undergo any instruction. And if this is difficult when one is trying to give new readers basic information about the library, it is doubly so when trying to persuade them to attend any more advanced instruction.

The Library Association working party report of 1949 recommended that an introduction to the library should be given to new students at the beginning of their first term. This is still generally accepted, and tours and introductory lectures are usually organized during 'freshers' week'. However, this is a time when new students are concerned primarily with settling in to their new surroundings, and for most the library comes low on their list of priorities. At this stage, the new undergraduates have little idea of what their library needs will be, and so

> "the first phase of orientation should be a generally pleasant introduction to the why and what of the library rather than how to use it."[43]

Instruction at this stage should be kept as simple and as general as possible, and in fact should be little more than "an inducement to visit and explore"[44] the library.

> "We have learned that it is futile to present anything beyond the most general information at an introductory lecture to new students. If we can convince them that we have a large library and a large staff ready to help them when they need help, we are laying a foundation for library patronage later."[45]

Some libraries try to rouse the student's interest in the library before his crowded first week of term. Pritchard reported in 1965[46] that the University of New Hampshire sent out pre-arrival instruction with the *Freshman Handbook*. This was in two parts: 'Before you arrive', which stressed the importance of knowing how to use the library, and 'Now that you are here', which dealt with the library's services and resources. His survey showed that 78% of the students said that they had read the library section of the handbook, and I gather that at New Hampshire this method of instruction is still being used.[47]

It seems that most libraries do not give instruction in the form of lectures at this introductory stage, although frequently the Librarian or one of his staff speaks to new students at the freshmen's conference or at some other suitable opportunity.

Instruction in the use of subject reference works

> "The image of library instruction as a single massive inoculation of freshmen against all further needs for information-search knowledge, appears to consciously or subconsciously condition the thinking of most faculty and students ... The idea of 'a' library lecture which dispenses all necessary knowledge must somehow be eliminated, if later specialized subject bibliographic instruction is to be accepted."[48]

This quotation pinpoints two of the major difficulties encountered when a course in library instruction is attempted. Library instruction should be a continuing process and not just something to be disposed of as quickly as possible. It is also necessary to convince the academic staff, firstly, that the students need such instruction and, secondly, that the library staff can do it better than they themselves can. Frequently this is no easy matter. A professor

> "is not likely to be impressed by the wish of the Librarian ... to teach him, his staff and students how to use the literature; he will argue that (a) he knows more about it than the library staff; (b) that his students will learn all they need to know from him, and (c) that the library should be concentrating on eliminating its backlog of cataloguing instead of wasting everybody's time."[49]

If a breach can once be made in this barrier and the academic staff can be convinced of the value of reader instruction and the competence of the librarians to give it, progress often seems to be quite rapid. Most of the librarians who have written on the subject, and all those to whom I have spoken, stress that without the co-operation of the academic staff a programme of library instruction is more often than not doomed to failure:

> "Any attempt to introduce a program of education in library use must have faculty backing."[50]

> "Library instruction prospers most when there is collaboration between the administration of the college and the library."[51]

It is of course impossible to generalize about the attitudes of members of the academic staff, but by and large those lecturers who have not been persuaded to let their students attend instruction classes frequently seem to be indifferent to them, and occasionally there are cases of active hostility. At Surrey, library instruction was encouraged from the beginning by Senate, and so it was easier there to persuade departments to make provision in their timetables for the students to attend courses given by the library. At City University, another technological university, some of the academic staff have taken the initiative in asking for instruction for their departments, but in other cases a lecturer has been convinced that the library staff have experience and expertise which he does not possess only when the Information Service has provided him with answers to problems which he could not solve himself. Without the backing of the academic staff, it would seem that the only way of ensuring a reasonable attendance is by making the library course compulsory. This was tried at Reading University, but because no other course at the University was compulsory, it generated resentment among the students.[52]

Once co-operation from the academic staff has been assured, it is essential to maintain close contact with them, so that the two crucial issues of timing and content can be arranged to the best advantage. Once again, the problem faced is that of student motivation. An undergraduate attends lectures in his main subjects because he knows that they are important and that eventually he will be examined in them. A lecturer must convince his students that knowing how to use a library is an essential part of their university education, and, moreover, directly relevant to their immediate needs. One successful method of bringing this home to the students is to time the library classes just before they are to start on a project. Co-operation with the lecturer is helpful, so that the classes may be held after the students have decided on their topics and just as they are starting to think seriously about them. If the librarian is briefed on the topics chosen he can use relevant examples and thus, one hopes, retain the attention of the students. This method was tried at Surrey University, where the lecturers advised their undergraduates to attend these classes, and in fact 80%-90% did so. There was some falloff as the course proceeded, but only of about 10%. Bradford University has had more difficulties with certain courses, and to combat a large drop in numbers after the first seminar it was decided to hold a concentrated one-day seminar. This successfully prevented students from disappearing after the first hour, but in spite of the use of a variety of teaching methods, including audio-visual aids, the staff concerned are sceptical about how much the students will have retained after this concentrated session. Southampton have also experimented with a whole-day seminar, but this has recently been cut down to a three-hour session after a survey conducted among the students who had attended suggested that boredom set in during the afternoon. The session comprises about one-and-a-half hours of teaching, followed by an hour's practical work and half an hour of teaching and questions at the end.

The content of library courses depends on the departments from which the students come and on the stage which they have reached in their courses. The ideal, of course, is that students should receive library instruction during all the years they spend at the university, but only in a few cases is this done at the moment, chiefly because of either lack of departmental support or lack of staff in the library to undertake such a heavy teaching load. The programme being evolved at Sussex University is similar to that already in operation at Surrey, as the member of staff in charge of instruction has recently moved there from Surrey. The 'orientation' at the beginning of the first year is general and tries not to impart too much information, as the students at this stage have no particular problems of library use. At the end of the first term or the beginning of the second, all the first-year undergraduates are given a talk on how best to use the library, and at this stage the catalogues, inter-library loans service, readers' advisory service, etc. are mentioned. City University also have a similar programme, and all these libraries have met with some success, perhaps because

> "by this time the students will know what their needs are and are more receptive to explanations of how the library can help to meet these needs."[53]

The emphasis at this stage is on finding the materials the students want (usually books on the reading list) and on what to do if they are unable to locate them.

For courses in subject work, the timing is even more important:

> "It is... necessary to ensure that a student is given a grounding in subject literature at a time when *we* know he needs it and *he* knows that he *wants* it."[54]

In other words, the student should have an idea of what he is looking for, and, perhaps, have experienced some difficulty in finding it. He will then be more ready to accept that there are things which he needs to be taught about finding information. It is in this part of the course where co-operation between the librarian and the lecturer on the academic staff needs to be closest. The importance of this from the point of view of timing and relevance has already been mentioned. It may also be possible in certain cases for the lecturer to take some part of the course himself, particularly where the subject content is high and where the lecturer has shown that he has sufficient bibliographic skills. It is important, however, that the overall running of the course should be in the hands of the library staff. Mr Crossley of Bradford University Library has stressed the dangers of turning too much of this teaching over to the academic staff:

> "I would go even further and state that very few teachers are *able* to give the necessary guidance in the use of subject literature, because they themselves have received no such initiation and make their own approaches to it haphazardly."[55]

Obviously the details of what should be taught vary enormously between the different disciplines, but the principles of any course in library use and exploitation are basically the same for any subject, even though the emphasis may change. Scrivener has summed these up as:

> "those traditional skills without which no student can make adequate use of his library. First, an understanding of library arrangements, physical, bibliographical and conceptual. Secondly, a knowledge of sources and of which will be appropriate in any given situation. Thirdly, the ability to interpret his own need so as to frame a relevant question. Fourthly, an awareness of search techniques including the ability to devise serviceable routines. Finally, the student needs skill in evaluating his sources and presenting his material."[56]

The length of the course itself depends usually upon such factors as the availability of staff for teaching and the willingness of the academic departments to allocate periods in the timetable. However, it seems to be the concensus of opinion that four hours is an absolute minimum in which to impart the essentials. In many cases the amount of instruction given is much more than this. Undergraduates at Surrey University receive on average a total of ten to twelve hours during their four years, and Bradford's full-day seminar has already been mentioned.

Almost every article written on the subject and every librarian to whom I have spoken seems to have a different view of the most appropriate methods of instruction. In many cases, the most suitable methods will be dictated by local factors such as the availability of rooms, the number of students, the subject, and by the personal preferences of the librarian himself. The traditional method of giving instruction in any subject is of course the lecture, and although it is at the moment unfashionable it does have the advantage that relatively large numbers of students can be taught at the same time. However,

> "the teaching of bibliographical tools in a conventional way can be very dull both for the teacher and the students,"[57]

and so it is important that the lecture should be made as

interesting and as varied as possible. Again the importance of the relevance of the content to the particular group of students cannot be overemphasized. The use of visual aids, such as slides or overhead transparencies, is also desirable, because students are more likely to be able to retain an image of the structure of literature if they can see it built up graphically before them. These aids can also illustrate sections of reference books when it would be impracticable for the students to see each book individually. Anyone who has given a formal lecture knows that feedback is generally minimal, with the result that misconceptions are difficult to clear up. For this reason seminars, where the groups are as small as possible, are probably more valuable in that the students are more likely to ask questions and reveal what they have not understood. The actual teaching should be concerned more with principles than with details. Nothing is more certain to lead to boredom than lists of "useful works on such-and-such a topic". This purely factual information is best given by means of a handout, which the students can annotate if necessary. (One of the handouts produced by Southampton University is to be published commercially in 1974.) The students can then concentrate more on what is being said, rather than have to scribble down titles, and the information which they are left with is also more likely to be accurate than any notes they have made themselves.

For there to be any hope that the students will retain much of what they have been told, it is important that some practical work is given. At Hatfield great importance is placed on the practical work, and in fact the lecturing time has been cut down as far as possible so that most of the real teaching is done individually as the students meet problems whilst doing the practical exercises:

> "Handling information is essentially a practical activity and each session of the course is in preparation for some practical work."[58]

The stress in these courses, it was emphasized to me, is not on learning how to use, say, *Science Citation Index* or *Chemical Abstracts* but on how to search for material. At City University the practical work is closely related to the work currently being done by the students in their own subjects, and in one department the practical work is marked jointly by the academic and library staff and counts as course work towards the students' final marks. Seeing this work as something essential to their degree leads to much greater co-operation from the students, and before the course starts each group is given a short preliminary test to see how much is already known, so that they will not be bored by being told what they are already familiar with. Southampton have found the case-study approach to be effective. Instead of unrelated questions requiring the use of periodicals or abstracts, questions are asked which are grouped round a specific problem such as the realization of the first laser, and students are to imagine that they are preparing a dissertation on such a topic. Indeed, if this approach can be used for a topic which the students actually do have to prepare, so much the better. Several libraries which give seminars attempt to evaluate their effectiveness. At Bradford a questionnaire was given to students who were completing a course of library instruction. They were asked whether the seminar achieved its aims of helping students to find and use reference books. Out of 28 replies, one said that it achieved these aims completely, 12 reasonably well, 14 satisfactorily, 1 not very well. When asked whether they would have liked more lecturing, 3 said 'yes' and 25 'no', but 18 would have liked more visual aids. As far as practical work was concerned, 4 felt that it was absolutely essential; 8 very useful; 9 useful; 5 not very useful; and 2 completely unnecessary.

Audio-visual aids

> "Giving library instruction to thousands or even hundreds of students is a monotonous and stultifying experience, consuming staff hours which could be put to a more fruitful use."[59]

This remark, made by an American librarian, illustrates some of the problems faced when trying to give instruction over a prolonged period to large numbers of students in small groups. In the United States, modern technology has been put to use in an attempt to solve some of these problems, and, as Table 2 indicates, these methods of mass communication are becoming increasingly widespread in British university libraries. The use of slides and overhead projection in conjunction with the traditional lecture has already been discussed. The most common types of audio-visual aid in use in British university libraries appear to be films, closed-circuit television and tape/slide presentations. Closed-circuit television is often used in connection with videotapes, so that the same lecture can be repeated to different groups of students. By exploiting the dramatic potential of movement, close-ups and graphics, a televised introduction to the library can be made much more palatable for the students than the normal lecture. It is, however, not cheap to produce and, as the time involved may be greater than that required for group tours, it ought to be possible to justify this type of instruction by demonstrating increased effectiveness. However, this is very difficult to measure, but when videotapes were used at Brunel the reaction was favourable.[60] Television has also been tried at Bishop Grosseteste College, Lincoln[61], and at Cardiff College of Education[62] and my survey indicates that it is at present in use at Brunel and Strathclyde Universities and the Polytechnic of Central London. Films have similar dramatic possibilities to those of television, and they have the added advantage that, compared with colour television, colour films are much easier and relatively cheaper to produce. But, whereas a videotape can be edited to bring it up to date with new developments, a film normally has to be completely remade. The expense, therefore, will be heavy. Although several libraries do use films in their instruction, it is perhaps significant that York no longer does so, and at Ulster the film made two years ago is already out of date.

The most promising method of mass instruction at present available seems to be the tape/slide presentation. This is

> "a sequence of projected pictures shown at appropriate points during a tape recording . . . The slides can either be changed by hand in response to audible signals on the tape recording or . . . the spoken commentary can be recorded on one track of the tape recording and synchronising pulses on another track; these pulses trigger an automatic slide projector so that slides are automatically shown at the correct points in the commentary."[63]

It is of course impossible to reproduce movement with these presentations (although this can be simulated by flashing several slides onto the screen in rapid succession), but the great advantage which they possess is flexibility. Slides can be taken out and replaced as necessary, and the problems of editing or remaking a tape recording are minimal. Some 41% of the libraries surveyed are already using tape/slide introductions and a further 14% are either planning or contemplating their use. They can also be used as guides to the literature of particular subjects or the use of major reference tools, and in 1970 a steering committee was set up to co-ordinate the co-operative production of such guides.

> "Subsequently the scheme was taken under the auspices of SCONUL and the steering committee is now the Working Party on Co-operative Production of Tape/Slide Guides to Library Services." [64]

By the end of 1972, 23 libraries had produced thirteen guides between them[65], covering such subjects as patents, Beilstein, *Chemical Abstracts*, the literature of sociology, and in the second round, now under way, thirteen more are to be produced, including case studies on the literature of various subjects, introductions to information retrieval and the structure of literature and 'How to use reference books'[66]. All of these guides are available for purchase, or for loan from the BLL. OSTI has awarded the University of Surrey a grant to

> "investigate procedures and models for the preparation and evaluation of tape/slide guides for use in information and library instruction."[67]

These presentations can be shown to large groups, either during freshers' week or in conjunction with library instruction classes, and they can also be made available for individual viewing. The University of Surrey has developed a double carrel for this purpose, and there are several more available commercially.

The increasing use of technological devices for instructing readers has been criticized by the Director of the Department of Audio Visual Communication of the British Medical Association[68], who described tape/slide presentations as "nothing more than a sophisticated lecture" and pointed out that unless they were used constantly they were a poor investment of capital. From a practical point of view they may be the most promising form of visual aid at present available, but experience suggests that few readers bother to use a tape/slide carrel. As with printed guides to the literature, one can appreciate their value when used as part of a course of library instruction or to give a general introduction to the library, but the reader who wishes to find a particular Parliamentary Paper will not normally want to sit through a twenty-minute tape/slide presentation on government publications. Perhaps the carrels would be more used if the presentation could give a brief answer to a specific question put to it, but here one is moving into the more advanced technological regions of the teaching machine, which, though widely used in American university libraries, has, as far as I know, not yet been much used in this country. The information stations at The Hatfield Polytechnic provide recorded instruction at the point of use, but they too provide more information than is needed immediately by the user. Perhaps, again, if they could provide specific answers to specific questions they would be made more use of than they are at present.

This chapter has concentrated on the mechanics of giving instruction to undergraduates. As far as graduate students are concerned, the principles are very much the same, although the 'orientation' part of the course need only be given to the graduates of other universities. The position may well in fact be easier with graduates, as the numbers are usually smaller, and, since graduate students have to search the literature in any case, the motivation to attend instruction ought to be stronger:

> "Graduate students are by far the most appreciative of library instruction. They know that they need help."[69]

Experience at Southampton, however, suggests that research students frequently do not appreciate the value of instruction until they have been engaged on their research for several years and have wasted such valuable time.[70]

Apart from these more formal methods of instruction, interest in the working of the library can be aroused by other public relations methods.[71] Quizzes are always popular, and in 1969 Cambridge University Library ran a bibliographic quiz giving book tokens as prizes. The questions involved not only using the library's catalogues (to put classmarks to titles) but also using reference books to elucidate abbreviations. The last few questions required no small amount of ingenuity in discovering, for example, who married George Walkerley, barber, of Wainfleet, All Saints, at Maltby in the Marsh on 1 September 1755. The last question required a short essay of 300 words:

> "Your Aunt Caroline, who has for many years been living in reduced circumstances, has sent you for Christmas an old book which, she says, she found in the attic. Its title page is reproduced herewith. Find out what you can about it and write a letter of thanks to Aunt Caroline."[72]

The book was *The whole booke of psalmes faithfully translated into English metre*, 1640; the winning answer was reprinted in *The Book Collector*.[73]

IV: THE INFORMATION OFFICER: INSTRUCTION OR INFORMATION?

The most effective method of instruction is of course to help the reader individually to help himself. Assisting with individual enquiries has been the task of the librarian at least since 1650, when John Durie wrote the quotation which begins this study. The mid-twentieth-century development of this concept has been the establishment of the library information officers.

The role of the librarian has always been in some measure to act as an intermediary between the books in his care and the reader wanting access to them. With the increasing complexity of information resources this role is becoming ever more necessary, indeed, as Mr M.B.Line points out:

> "I doubt very much if a system can be highly efficient and sophisticated *and* directly usable by ordinary researchers."[74]

Many university libraries have, therefore, established services known variously as Readers' Advisory Services, Information Services, etc. Simple questions about the arrangement of books in the library, or quick-reference questions are frequently answered from the issue desk or some other suitable point. The answering of such questions and the advice to readers on which reference sources to use are services which university libraries have provided for many years. The enquiries which are referred to the information officer/readers' adviser are generally, however, more complex.

Between 1970 and 1973 OSTI financed information services in six academic libraries: Birmingham, Salford, Strathclyde and Sussex Universities, University College Cardiff and Imperial College London. It was envisaged that

> "each information officer will be responsible for the education and training of staff and postgraduate students in the use of information resources and will promote effective use of information services."[75]

It has been estimated[76] that in 1972 there were about 35 information officers in British university libraries, offering either brief advice or

> "a complete range of services by providing, for example, personal bibliographies, translations, guides, current awareness services, and even lecture courses or teaching programmes for nearly all members of the staff and student body."[77]

This view of the service which an academic library should provide goes well beyond the traditional reference function and approaches that found in special libraries in industry. In such libraries, when a scientist requires information he knows that there is a librarian or information scientist available to search for it for him. In university libraries the tradition has been to tell the reader where to find the bibliographies or reference works which he needs, and then to leave him to do the work himself.

However, with the increase in information services the question has arisen as to how much help the library should provide. Should it provide *information* (everything the user requires) or *instruction* (showing the reader how to find the information for himself)? It has been suggested that at present readers do not know how far the library is able or willing to help its readers:

> "library patrons have been left uncertain as to just how much and what kind of instruction or information is offered, or how much they may justifiably expect of a librarian."[78]

The same writer suggests, furthermore, that the provision of instruction is merely a secondary goal:

> "The library's essential educational obligation consists in something quite different from teaching its use ... The library supplies the most accurate and complete information—the sources for education."[79]

In Dr Urquhart's opinion, however, the educative function of the library consists not in providing information so much as in developing user independence. The information officers

> "should be less concerned with attempting to carry out literature searches for their clients and be more concerned with showing them how to search efficiently for themselves."[80]

A survey carried out at Bath after the establishment of an information service for the social sciences revealed that academics in these subjects were generally in favour of the 'information' approach: "clients would rather be served than learn to help themselves"[81]. The main reasons given for the value of the service were: (1) time—most academics did not have time to conduct their own searches, (2) wider coverage—the information officers knew of more sources than the academics (3) division of labour—many academics felt that literature searching was more the job of the information officer than of the academic, (4) volume of material—the information officer acted as a filtering device.[82]

Given a sufficient number of information officers, the provision of information will frequently save the time of the academic. However, it is likely that no library, however well staffed, will be able to meet all the demand all the time, and so if the academic does not know how to search for himself he may have to submit to a considerable delay in obtaining the information he needs. Furthermore, he may want information on a confidential or personal matter and not wish to expose this to the eyes of a third party. Again, it would be more useful for him if he were able to search for himself:

> "Training in bibliographical research is for the user's convenience, economy, privacy, satisfaction and education."[83]

Should the library provide information, then, apart from the answers to quick-reference enquiries? Even those most in favour of the 'information' approach would generally distinguish between the service provided for the academic staff and that for research and undergraduate students. It is widely accepted that literature searching is part of the training of the student, particularly of the research student, and indeed that

> "the thesis is partly an exercise in assembling all the relevant information on a problem".[84]

The information officer should certainly assist the research student, and it is here that individual instruction comes into its own. One would hope that the student has had some formal tuition in the techniques of literature searching and in the structure of literature, but by seeing how the librarian undertakes a search he will be more equipped to proceed further on his own. This technique is used at Loughborough:

> "An hour can be very well spent as such instruction gives the student a real opportunity to assess the value of information for himself. A popular course of action is the provision of a few good references by the information officer during the initial stages of the search and the student taking over from there. This approach is very much appreciated by the students, particularly the postgraduates."[85]

As far as service to the academic staff is concerned, this combination of 'instruction' and 'information' is probably the clue to the most effective use of the information officer. Scientists are generally more willing to entrust their literature searching to an information scientist or librarian than are academics in the humanities. Pressure for an information service in the humanities is virtually non-existent at the moment, because most academics in this field, it would seem, do much of their research from references already obtained or by browsing. The information officer's main task here is that of instruction—helping the scholar towards a more systematic approach to the literature of his subject. Although this appears to be the situation at the present time, so little research has been done on the information needs of the humanities that

> "if [information] services were introduced a volume of latent demand might be unleashed".[86]

Even with scientists the position is not entirely straightforward. True, the information officer is probably best fitted to provide factual information on a specific problem or, knowing the individual scientists' interests, to give a current-awareness service. But scientists also have to search for new ideas, and here it is the individual himself who must decide what is relevant—he cannot pass this task on to a librarian. The scientist must be able to search efficiently for himself, and so he requires not only information but also instruction. At the City University the intention of the information service

> "is not in any way to spoon-feed or supplant normal academic processes but rather to promote what has been termed 'library-user independence' at a time when the sheer mass of published and unpublished information is reaching such daunting and unmanageable proportions." [87]

Although the information section there does give some formal instruction, it

> "maintains its original principle of service to the individual and almost certainly the aspect of its work most appreciated is the combination of *ad hoc* literature searching with instruction."[88]

It is this combination of 'instruction' with 'information' which seems to be the most promising. When an academic requires some information or a number of references on a specific topic, the information officer is generally more qualified to provide it efficiently than is the academic himself:

> "If a necessary job can be done far more efficiently by someone else, you don't do it yourself."[89]

Besides, experience at Bath[90] and elsewhere indicates that researchers are less likely to put as much effort into a literature search as is an information officer, and thus a search carried out by him will be more thorough and, presumably, more valuable. However, when the researcher is outside his own field, or less able to specify the subject of his search, he and the information officer may have to work together, or he may have to conduct a preliminary survey himself to establish more closely what exactly he is looking for. The aim, therefore,

> "should be not to make him either fully independent or slavishly dependent, but to establish a balance whereby both he and the information officer carry out the activities for which they are best fitted."[91]

The information officer's task, then, is, as Durie said, to "propose to others [the publick stock of Learning] in the waie which may be most useful unto all". This, indeed, is the task of all librarians, particularly those concerned with the instruction of readers. With the increasing stress being laid on reader services it is likely that this is an aspect of librarianship which will continue to expand rapidly. As librarians become more adept at instruction and develop more sophisticated teaching methods it is to be hoped that this instruction will become more effective and more widely accepted than it is at present, so that a similar survey to this, carried out in ten years' time, may perhaps be able to report that regular seminars, fitted into the academic timetable, are as much a part of the academic library scene as are the printed handbooks at the moment.

V: BIBLIOGRAPHY

References

1. MEWS, H. Library instruction concerns people. *Library Association Record,* 72 (1), 1970: 8–10, p. 8.

2. POTTS, H.E. Instruction in bibliographical technique for university students. *ASSOCIATION OF SPECIAL LIBRARIES AND INFORMATION BUREAUX. Report of proceedings of the Third Conference.* London: ASLIB, 1926.

3. TIDMARSH, M.N. Instruction in the use of academic libraries. *SAUNDERS, W.L. ed. University and research library studies.* Oxford: Pergamon, 1968, p. 39.

4. CROSSLEY, C. Education in literature and library use. *The Library World,* 71 (839), 1970: 340–347, p. 340.

5. RICHNELL, D.T. The Hale Committee report and instruction in the use of libraries. *Library Association Record,* 68 (10), 1966: 357–361, p. 358.

6. POWER, E. Instruction in the use of books and libraries; preliminary report to the International Association of Technical University Libraries. *Libri,* 14 (3), 1964: 253–263, p. 253.

7. MELUM, V.V. Library instruction in a university. *Illinois Libraries,* 51 (6), 1969: 511–521, p. 511.

8. UNIVERSITY GRANTS COMMITTEE. Report of the Committee on University Teaching Methods. (Hale Committee.) London: HMSO, 1964.

9. ASSOCIATION OF UNIVERSITY TEACHERS. The university library. [London?] : AUT, 1964, p. 5.

10. UNIVERSITY GRANTS COMMITTEE. Report of the Committee on Libraries. (Parry Report.) London: HMSO, 1967, para. 504.

11. SCRIVENER, J.E. Instruction in library use: the persisting problem. *Australian Academic and Research Libraries, 3* (2), 1972: 87–119, p. 101.

12. UNIVERSITY GRANTS COMMITTEE. Report of the Committee on Libraries. London: HMSO, 1967, Appendix 3.

13. UNIVERSITY GRANTS COMMITTEE. *ibid.*, Appendix 5.

14. UNIVERSITY GRANTS COMMITTEE. *ibid.*, para. 456.

15. UNIVERSITY GRANTS COMMITTEE. *ibid.*, para. 448.

16. WOOD, D.N. Instruction in the use of scientific and technical literature. *Library Association Record, 70* (1), 1968: 13.

17. WOOD, D.N. *and* BARR, K.P. Courses on the structure and use of scientific literature. *Journal of Documentation, 22* (1), 1966: 22–32, p. 23.

18. LIBRARY ASSOCIATION. UNIVERSITY AND RESEARCH SECTION. Working party on instruction in the use of libraries and in bibliography at the universities.. Report. *Library Association Record, 51* (5), 1949: 149–150.

19. CAREY, R.J.P. The teaching and tutorial activities of librarians with students not training for the library profession. FLA Thesis, 1966.

20. CAREY, R.J.P. *ibid.*, p. 14.

21. MALTBY, A. UK catalogue use survey: a report. London: The Library Association, 1973, p. 10.

22. WENDT, P.R. Programmed instruction for library orientation. *Illinois Libraries, 45* (2), 1963: 72–77, p. 73.

23. GRIFFITH, A.B. Library handbook standards. *Wilson Library Bulletin, 39* (6), 1965: 475–477, p. 476.

24. BRYNMOR JONES LIBRARY. Readers' guide, 1972. Hull: University of Hull, 1972, p. 10.

25. BRYNMOR JONES LIBRARY. *ibid.*, p. 23.

26. UNIVERSITY OF BRADFORD. Know your library. 4th ed. Bradford: University of Bradford, 1972, p. 15.

27. HUTCHINS, W.J., PARGETER, L.J., SAUNDERS, W.L. The language barrier: a study in depth of the place of foreign language materials in the research activity of an academic community. Sheffield: University of Sheffield Postgraduate School of Librarianship and Information Science, 1971, p. 207.

28. DALLEY, S.P. A study of printed guides to catalogues for the use of readers in British university libraries. A study submitted in partial fulfilment of the requirements for the degree of M.A. in librarianship. University of Sheffield, 1971.

29. DALLEY, S.P. *ibid.*, p. 42.

30. MALTBY, A. Measuring catalogue utility. *Journal of Librarianship*, 3 (3), 1971: 180–189, p. 184.

31. PALMER, M.C. Library instruction at Southern Illinois University, Edwardsville. *Drexel Library Quarterly*, 7 (3/4), 1971: 255–276, p. 269.

32. LIBRARY ASSOCIATION. Libraries in the new polytechnics. *Library Association Record*, 70 (9), 1968: 240–242, p. 241.

33. CAREY, R.J.P. A systems approach to exploiting library resources: some experimental factors and problems. *CAREY, R.J.P. ed. Exploitation of library resources; a systems approach. A workshop at The Hatfield Polytechnic, April 22nd, 1972.* Hatfield: The Hatfield Polytechnic, 1972, p. 13.

34. CHILDS, N. Do students have to use libraries? *The Times Higher Education Supplement*, No. 85, 1 June 1973, p. 24.

35. CAREY, R.J.P. A systems approach to exploitation. *New Library World*, 73 (865), 1972: 347—349, p. 347.

36. KNIGHT, C.J. Colour coding and the provision of non-print material at the University of Bath Library. *CAREY, R.J.P. ed. Exploitation of library resources; a systems approach. A workshop at The Hatfield Polytechnic, April 22nd, 1972*. Hatfield: The Hatfield Polytechnic, 1972, pp. 20—21.

37. PIPE, C.P. The university and its library. *Library Association Record*, 70 (9), 1968: 180—181, p. 181.

38. LINE, M.B. Information services in academic libraries. *Educating the library user. IATUL: Proceedings of the fourth triennial meeting, 1970*. Loughborough: University of Technology Library, 1970, p. B-2.

39. RONKIN, R.R. A self-guided library tour for the biosciences. *College and Research Libraries*, 28 (3), 1967: 217—218, p. 218.

40. RONKIN, R.R. *op. cit.*

41. SCRIVENER, J.E. *op. cit.*, p. 105.

42. KIRTLEY, E.B. Private communication. *Quoted in:* ROBERTS, N. University libraries. *Library Association Record*, 75 (3), 1973: 48—50, pp. 48—49.

43. HARTZ, F.R. Freshman library orientation: a need for new approaches. *College and Research Libraries*, 26 (3), 1965: 227—231, p. 228.

44. HACKMANN, M. Proposal for a program of library instruction. *Drexel Library Quarterly*, 7 (3/4), 1971: 299—308, p. 307.

45. MELUM, V.V. *op. cit.*, p. 520.

46. PRITCHARD, H. Pre-arrival instruction for college students. *College and Research Libraries*, 26 (4), 1965: 321.

47. PRITCHARD, H. Personal communication, 1973.

48. PALMER, M.C. *op. cit.* pp. 257 *and* 261.

49. MACKENZIE, A.G. Reader instruction in modern universities. *Aslib Proceedings, 21* (7), 1969: 271–279, p. 276.

50. BECHTEL, J.M. A possible contribution of the Library-College idea to modern education. *Drexel Library Quarterly, 7* (3/4), 1971: 189–201, p. 194.

51. WÓJCIK, M. Academic library instruction. *College and Research Libraries, 26* (5), 1965: 399–400, p. 399.

52. MEWS, H. Library instruction to students at the University of Reading. *Education Libraries Bulletin*, No. 32, 1968: 24–34.

53. WILL, L.D. Finding information: a course for physics students. *Physics Bulletin, 23*, 1972: 539–540, p. 539.

54. CROSSLEY, C. *op. cit.*, p. 345.

55. CROSSLEY, C. *ibid.*, p. 341.

56. SCRIVENER, J.E. *op. cit.*, p. 103.

57. BRISTOW, T. Instruction or induction: the human approach to student involvement in library materials. *Seminar on human aspects of library instruction... held at the University of Reading.* Cardiff: SCONUL [1970?], p. 11.

58. CAREY, R.J.P. A systems approach to exploitation. *New Library World, 73* (865), 1972: 347–349, p. 348.

59. HARTZ, F.R. *op. cit.*, p. 229.

60. WYATT, R.W.P. The production of video-tapes for library instruction—an account of experience at Brunel University. *Educating the library user. IATUL: Proceedings of the fourth triennial meeting, 1970.* Loughborough: University of Technology Library, 1970, pp. L1–L5.

61. DEAN, E.B. Television in the service of the library. *Library Association Record*, 71 (2), 1969: 36—38.

62. OWEN, E.M. Closed-circuit television in the library. *Education Libraries Bulletin*, No. 27, 1966: 24—27.

63. EARNSHAW, F. *ed.* Tape/slide presentations: recommended procedures. Cardiff: SCONUL, 1973, p. 2.

64. BARR, K.P. SCONUL tape/slide presentations for library instruction. *CAREY, R.J.P. ed. Exploitation of library resources; a systems approach. A workshop at The Hatfield Polytechnic, April 22nd, 1972.* Hatfield: The Hatfield Polytechnic, 1972, p. 19.

65. OFFICE FOR SCIENTIFIC AND TECHNICAL INFORMATION. Evaluation of tape/slide guides to library and information services. *OSTI Newsletter*, 1972: 4, p. 9.

66. STANDING CONFERENCE OF NATIONAL AND UNIVERSITY LIBRARIES. SCONUL Working Group on Tape/Slide Guides to Library Services. *Newsletter*, May 1973.

67. OFFICE FOR SCIENTIFIC AND TECHNICAL INFORMATION. *ibid.*

68. ENGEL, C.E. Letter to *The Times Higher Education Supplement*, No. 81, 4 May 1973, p. 12.

69. MELUM, V.V. Library orientation in the college and university. *Wilson Library Bulletin*, 46 (1), 1971: 59—66, p. 63.

70. LINE, M.B. *op. cit.*

71. MEWS, H. Reader instruction in colleges and universities: an introductory handbook. London: Bingley, 1972.

72. CAMBRIDGE UNIVERSITY LIBRARY. Duplicated sheet of questions, 1969.

73. THE BOOK COLLECTOR. Article. *18*, 1969: 382—383.

74. LINE, M.B. *op. cit.*, p. B-4.

75. DEPARTMENT OF EDUCATION AND SCIENCE. OSTI: the first five years. London: HMSO, 1971, p. 26.

76. MYATT, A.G. Information officers in university libraries. *NLL Review*, 2 (3), 1972: 85—88.

77. MYATT, A.G. *ibid.*, p. 85.

78. SCHILLER, A.R. Reference service: instruction or information. *Library Quarterly, 35* (1), 1965: 52—60, p. 59.

79. SCHILLER, A.R. *ibid.*, p. 57.

80. URQUHART, D.J. Developing user independence. *Aslib Proceedings, 18* (12) 1966: 351—356, p. 352.

81. BATH UNIVERSITY LIBRARY. Experimental information service in the social sciences, 1969—1971: final report. Bath: Bath University Library, 1972, p. 73.

82. BATH UNIVERSITY LIBRARY. *ibid.*, p. 75.

83. STOBART, R.A. Use of library — or GCE Bibliographical Knowledge. *The Library World, 70* (819), 1968: 55—63, p. 55.

84. DANNATT, R.J. Books, information and research; libraries for technological universities. *Minerva,* 5 (2), 1967: 209—226, p. 224.

85. RHODES, R.G. *and* EVANS, A.J. The educational role of the university library and the provision of information services. Unpublished paper. University of Technology, Loughborough.

86. LINE, M.B. Information services in university libraries. *Journal of Librarianship, 1* (4), 1969: 211—224, p. 213.

87. ENRIGHT, B.J. The University Library: key to education and communication. *Quest: the Journal of the City University, London*, No. 4, 1968: 6—14, p. 8.

88. CORNEY, E. The information service in practice: an experiment at the City University Library. *Journal of Librarianship, 1* (4), 1969: 225—235, p. 233.

89. LINE, M.B. Information services in university libraries. *Journal of Librarianship, 1* (4), 1969: 211—224, p. 219.

90. BATH UNIVERSITY LIBRARY. *op. cit.*

91. LINE, M.B. Information services in academic libraries. *Educating the library user. IATUL: Proceedings of the fourth triennial meeting, 1970.* Loughborough: University of Technology Library, 1970, p. B-5.

Other Works Used

APTED, S.M. The university library user and his information needs. Essay submitted in part fulfilment of requirements for M.A. (Librarianship), University of London, 1970.

ASHWORTH, W. The information officer in the university library. *Library Association Record, 41* (12), 1939: 583—584.

BESWICK, N.W. "Library-college" re-visited. *Library Association Record, 72* (7), 1970: 148—149.

CAREY, R.J.P. Library instruction in colleges and universities of Britain. *Library Association Record, 70* (3), 1968: 66—70.

CAREY, R.J.P. A technical information course for engineering and science students at Hatfield College of Technology. *Library Association Record, 66* (1), 1964: 14—20.

CHESSHYRE, H.A. *and* HILLS, P.J. Evaluation of student response to a library instruction trials programme using audio-visual aids. *Educating the library user. IATUL: Proceedings of the fourth triennial meeting, 1970.* Loughborough: University of Technology Library, 1970, K1–K11.

CROSSLEY, C.A. Tuition in the use of the library and of subject literature in the University of Bradford. *Journal of Documentation, 24* (2), 1968: 91–97.

EARNSHAW, F. Co-operative production of tape/slide guides to library services. *Library Association Record, 73* (10), 1971: 192–193.

GENUNG, H. Can machines teach the use of the library? *College and Research Libraries, 28* (1), 1967: 25–30.

GORE, D. Anachronistic wizard: the college reference librarian. *Library Journal, 89,* 1964: 1690.

HOWISON, B.C. Simulated literature searches. *Drexel Library Quarterly, 7* (3/4), 1971: 309–320.

HUTTON, R.S. Training students in the use of libraries. *Universities Quarterly, 4* (4), 1950: 389–392.

KENNEDY, J.R. *et al.* Course-related library instruction; a case study of the English and biology departments at Earlham College. *Drexel Library Quarterly, 7* (3/4), 1971: 277–297.

LARSON, T.E. The public onslaught: a survey of user orientation methods. *RQ, 8* (3), 1969: 182–187.

LINE, M.B. University libraries and the information needs of the researcher: a provider's view. *Aslib Proceedings, 18* (7), 1966: 178–184.

LINE, M.B. *and* TIDMARSH, M. Student attitudes to the university library: a second survey at Southampton University. *Journal of Documentation, 22* (2), 1966: 123–135.

MACKENNA, R.C. Instruction in the use of libraries: a university library problem. *Journal of Documentation, 11* (2), 1955: 65–72.

MACKENZIE, A.G. A new university prospect. III: Service, staffing, equipment and methods in the new university. *Aslib Proceedings, 17* (4), 1965: 112–120.

MARSHALL, A.P. Library outreach: the program at Eastern Michigan University. *Drexel Library Quarterly, 7* (3/4), 1971: 347–350.

MELUM, V.V. 1971 survey of library orientation and instruction programs. *Drexel Library Quarterly, 7* (3/4), 1971: 225–253.

MEWS, H. Teaching the use of books and libraries, with particular reference to academic libraries. *WHATLEY, H.A. ed. British librarianship and information science, 1966–1970.* London: The Library Association, 1972, 601–609.

MYATT, A.G. Experience in educating the user at the NLL. *Educating the library user. IATUL: Proceedings of the fourth triennial meeting, 1970.* Loughborough: University of Technology Library, 1970, F1–F17.

PARRY, T. University libraries and the future. *Library Association Record, 70* (9), 1968: 225–229.

PHIPPS, B.H. Library instruction for the undergraduate. *College and Research Libraries, 29* (5), 1968: 411–423.

PUGH, L.C. Library instruction programmes for undergraduates: historical development and current practice. *The Library World, 71* (837), 1970: 267–273.

REVILL, D.H. Teaching methods in the library: a survey from the educational point of view. *The Library World, 71* (836), 1970: 243–248.

What catalogue? *Catalogue and Index*, No. 8, 1970: 1.

Additional References

Three further items which have appeared since the submission of this study are particularly relevant:

CAREY, R.J.P. Library guiding: a systems approach to exploitation. London: Bingley, 1973.

DUREY, P. Printed guides to university libraries. *Australian Academic and Research Libraries,* 4 (2), 1973: 85—90.

HALL, J. Survey of information services provided by British university libraries, 1973. Library profiles. Sheffield: Sheffield University Library, 1973.

SEP 2 1976

Z
675
U5
F69

Z
675
U5
769